Social Clustering: Paradigm of Trust

BALTISCHE STUDIEN ZUR ERZIEHUNGS- UND SOZIALWISSENSCHAFT

BALTIC STUDIES IN EDUCATIONAL AND SOCIAL SCIENCES

Herausgegeben von Gerd-Bodo von Carlsburg, Natalija Mažeikienė und Airi Liimets

Edited by Gerd-Bodo von Carlsburg, Natalija Mažeikienė and Airi Liimets

BAND 35

Zu Qualitätssicherung und Peer Review der vorliegenden Publikation

Die Qualität der in dieser Reihe erscheinenden Arbeiten wird vor der Publikation durch die Herausgeber der Reihe geprüft.

Notes on the quality assurance and peer review of this publication

Prior to publication, the quality of the work published in this series is reviewed by editors of the series.

Giedrė Kvieskienė / Vytautas Kvieska /
Gerd-Bodo von Carlsburg

Social Clustering: Paradigm of Trust

PETER LANG

Bibliographic Information published by the Deutsche Nationalbibliothek
The Deutsche Nationalbibliothek lists this publication in the Deutsche Nationalbibliografie; detailed bibliographic data is available online at http://dnb.d-nb.de.

Library of Congress Cataloging-in-Publication Data
A CIP catalog record for this book has been applied for at the Library of Congress.

This book was printed with the support of Vytautas Magnus University.

Cover image: ©Vytautas Kvieska

ISSN 1434-8748
ISBN 978-3-631-81726-1 (Print)
E-ISBN 978-3-631-83626-2 (E-PDF)
E-ISBN 978-3-631-83627-9 (EPUB)
E-ISBN 978-3-631-83628-6 (MOBI)
DOI 10.3726/b17624

Peter Lang – Berlin · Bern · Bruxelles · New York · Oxford · Warszawa · Wien

This publication has been peer reviewed.

www.peterlang.com

Preface

The monograph consists of four parts. The first part Modeling Social Economy and Social Clustering comprises two parts: Community-Based Social Industries for Creative Economy and Smart Education Indicators. The second part of the monograph is devoted to examining the issues of Social Economy and Civic Engagement. The third part of the book is called Social Mediation and Communication. Fourth part deals with the problems of the Region Policy: The regional level the content touches upon is more general points of new economy models, smart education and analyses of success case. Social clustering theories, success stories and models of the sustainability algorithms for smart education and social partnership are analyzed in this monograph. Concepts that are conceptualized in articles, studies, monographs and research reports are associated with smart education, child welfare and positive socialization. This concept is also widely used in Lithuania for the preparation of national and international strategies, concepts, normative documents and projects between both public and non-governmental organizations. The phenomenon of social clustering development has not been analyzed sufficiently, and its development has been hampered both in Lithuania and in other countries. By analyzing a smart education scenario, both the scientists and stakeholders experiment to acquire answers about empowering public, private, research institutions and NGOs for daily networking, social clustering, developing social capital and public welfare.

Approved by the Board of the Education Academy, Vilnius and Kaunas, Vytautas Magnus University, December 3, 2019.

Reviewed:

Prof. Dr. Vilija Targamadzė, Vilnius University,

Prof. Dr. Giedrė Misiūnienė, Vytautas Magnus University,

Assoc. Prof. Dr. Arvydas Liepuonius, Vytautas Magnus University.

Contents

Concepts

Advertising – "means information disseminated in any form and by any means and relating to a person's commercial and economic, financial or professional activities, where it promotes the purchase of goods or use of services, including the purchase of immovable property and the takeover of property rights and obligations" (Law on Advertising of the Republic of Lithuania). Advertising audience – potential buyers, clients or spectators to whom advertising is devoted to.

Associations/confederations – social partnership networks of professionals, non-governmental organizations, employers and other civil organizations in the public and private sectors formed/self-formed in pursuance of negotiations, consultations, lobbying and encouraging social responsibility of all sectors (Kvieskienė, Kvieska, 2012) adopting the term of Association of Directors of Social Service from Martin Thomas and John Pierson „Socialinio darbo žodynas" (2010, p. 30).

Assisted living technologies – consist of software and appliances that help elderly people live independently and improve their standard of living. For example, Telecare solutions consist of using monitors and censors that enable patients to receive medical care at home. Research suggests that the use of such technology across the National Health Service (NHS) can result in £1 billion of annual saving.[1]

Brand – not only the mark or logo of the product or the whole company. It is a broader term, which exists only in the head of a person. A brand is not only a company logo, it, according to image branding "guru" J. Grant, is a "collection of (strategic) cultural ideas." This is one of the main organization's marketing communication goals – development and increase of trademark value. A brand is related to all associations in the head of a person when one or the other word is mentioned when a mark or even color is noticed.

Business incubator – a public, private or civic institution/organization operating in a territory of a region, or one or more municipalities that under preferential

---

1 Examples are drawn from BIS (2013), The Smart City Market: Opportunities for the UK, London: Department for Business Innovation and Skills. Internet access: https://assets.publishing.service.gov.uk/government/uploads/system/uploads/attachment_data/file/246019/bis-13-1209-smart-cities-background-paper-digital.pdf

conditions rent premises and technical and office equipment and provide business information and consulting and training services to tenant companies.

Community – hall of shared affairs that joins people or organizations, as agreement for the shared activity is based on voluntary and mutual assistance and self-regulation and self-governance principles, which guarantees safe working and living environment as well as spiritual health[2].

Community – an organized group of people who share a common task and often-common homeland and depend on each other's success in seeking the survival and well-being (G. Kvieskienė, 2003).

Community abilities – qualitative lifestyle expression of the community, which shows how shared agreement of persons realized through the community activity that in the social-political environment manifests itself by the community, community organization[3].

Discourse – the unit of articulate practices characterized by a certain unifying attribute. For example, the modern discourse.

Ecosystem – a system or a group of interconnected elements, formed by the interaction of a community of organizations and individuals with their environment.

Entrepreneur - someone who starts his or her own business, especially when this involves seeing a new opportunity.

Institution – social structure, which is formed and maintained by various normative, regulative and cultural cognitive institutional elements.

Institutional powers – entirety of acting factors in the environments that are upheld by institutionalized practices, norms, regulations, standards, values and other aspects inherent by institutions.

Incubation – rent of premises, technical and office equipment as well as provision of business information, consultation and training on preferential terms to incubated entities.[4]

---

2 Nefas S. (2007), Funkcionali vietos bendruomenė Lietuvos kaimuose ir miesteliuose. Daktaro disertacija. Socialiniai mokslai, vadyba ir administravimas (S03). Vilnius: Mykolo Romerio universitetas.

3 NVO sektoriaus verslo galimybių formavimo vystant socialinę ekonomiką Pasvalio rajono kaimo vietovėse koncepcija.1013, Pasvalys.

4 Incubation term. Internet access: http://inkubatorius.lt/?page_id=42

Indicators – measuring rates validated in the global or national indices.

Innovation – installation of new or substantially improved products (goods or services) or processes, new marketing or organizational methods in the business practice, organization or while developing external relations.[5]

Innovation system – the entirety of interrelated institutions and organizations and their interaction relations and measures allowing expanding new technologies and innovations.

Integrated marketing communication (IMC) – in 1989 the variant of American Marketing Association claims that IMC is a special planning conception of marketing communication that aims at creating marketing communication action plan in which diverse elements (commercial, direct marketing, sales promotion and public relations) would deliver a clear, consistent and stable communicative message of the maximum effect (Christensen, Torp, Firat, 2005), (Kotler P., 2003) and also supplements that IMC demands for integration and coordination of different communication channels.

Interactive marketing – is an interactive process developed in the information technology space, inside the company by establishing and building relations with the consumer on purpose to present and sell a product.

The ISO standardization process – an International Standard is the result of an agreement between ISO member bodies. It may be used as such, or may be implemented through incorporation in national standards of different countries.[6]

Local community – a particular social group, which is based on small, specific geographical location; the residents are linked by the specific relationship (Poviliūnas A., 2003, p. 71).

Life quality – agreed community lifestyle standard based on the public spirit and national identity in the contexts of worldwide experience and the human destination.

Multifunctional design – this is ecosystem projection of environment and activities oriented to the needs of person and the community with the purpose of their interaction and positive synergy (Kvieskienė, Bardauskienė, 2014; Šliužienė, 2012).

---

5 Internet access: https://www.google.lt/#q=socialin%C4%97s+inovacijos

6 The ISO standardization process. Internet access: http://www.isotc292online.org/projects/iso-process/

Positive socialization – policy of positive actions (scenario of positive socialization) based on which certain persons or groups of persons adjust the behavior of socialization subjects seeking to preserve positive components of the culture and prevent consequences of elements that are unacceptable in this culture (Kvieskienė, G., 2007).

Positioning – the value place of the brand (company and product) in the consumer's consciousness, which company seeks to reach and its benefits. The aim is to force the consumer think about these benefits, separating the concrete brand from the competitors.[7]

Provision of integrated help – provision of educational assistance, social support, health-care services to children and their parents (guardians) living within the territory of the municipality.

Public, private and NGO partnership – describes the validated agreements between the public and private sectors in the development of public infrastructure, taking into account the real needs of the community or the provision of other related services.

Public services for small- and medium-sized enterprises (hereinafter – the public service for business) – information, advice, training, continuous education or retraining, methodological and other services to business start-up, business planning, fundraising, financial management, staff management, innovation activation, cooperation, technology development and transfer, other business-related issues, provision in various communication tools on preferential terms to small- and medium-sized business entities and (or) natural persons who intend to start a business, as well as the facility, hardware and office equipment rental on preferential terms and practical help for small- and medium-sized enterprises renting these premises.

Regional development – a broad term but can be seen as a general effort to reduce regional disparities by supporting (employment and wealth-generating) economic activities in regions. In the past, regional development policy tended to try to achieve these objectives by means of large-scale infrastructure development and by attracting inward investment.[8]

7 Internet access: http://pcode.eu/lt/menu1/marketingas/marketingo-zinynas/pozicionavimas/

8 Regional development policy. In: http://www.oecd.org/cfe/regional-policy/regionaldevelopment.htm

Regional development is a complex concept and can be described as a balanced effort to strengthen cross-sectoral social partnerships, which help to reduce regional disparities by initiating entrepreneurship, free market areas, supporting projects that develop creativity, employment and prosperity and stimulate economic activity in concrete region.

Smart cities – a networked culture-driven process that fosters civic and NGO engagement, and the formation and empowerment of local communities with shared responsibility and decision-making, promoting modernization of public spaces and social capital through the development of digital technologies, artificial intelligence, that make cities more viable, more resilient and responsive to challenges.

Smart city – a safe, modern, open, encouraging civic initiative, resource efficient and a culture of sharing. The smart city is characterized by the promotion of modern technologies, an open internet connection, the modernization of public spaces and public institutions, their opening to the citizen and the application of the universal design principle. Smart initiatives in the city can range from sharing bicycles, cars to multifunctional centers set up in existing or adaptable public institutions, and the decisions taken must be open and inclusive of all stakeholders: public, private, academia and civil society.

Social capital – according to the OECD, is a networked culture based on values and mutual understanding, facilitating cooperation within and between social groups. In this definition, network culture is understood as a prerequisite for building and developing trust and social relationships between groups or individuals. It is about networks of friends, family and former colleagues who share common goals and interests.

To sum up, networking culture and fostering mutual understanding build trust, foster a culture of sharing and enable people to work and create together for prosperity.

Smart initiatives – this can be small or large scale and range from bike sharing to integrated operations centers, while the number of stakeholders involved can be so numerous (city councils, technology companies, citizens, universities and charities) that the vision for a smart city changes depending on whom you ask. As a result, the concept is poorly defined and understood, and is at risk of sitting alongside other well-used but rarely defined notions like "liveability" and "sustainability."[9]

9 Campbell M (2014), Smart Cities? Not for me!, blog post published on 28 February, 2014 on http://professormikecampbell.wordpress.com and Townsend A (2013),

Social business (social enterprise) – a kind of business with the main social or environmental purpose and not the goal of increasing the profit for shareholders. Accordingly, the earned profit is reinvested back to business or used for charitable and social activities. Social business may involve operational principles and form of the activity. As a process, it manifests itself by appliance of business principle in order to solve social and environmental problems. As activity form, it is created in order to solve social problems, it reinvests profit instead of paying its partners and at least half of its profit comes from the sale of services or goods. Some social businesses are characterized by greater involvement of employees and equal ownership and decision-making rights division among partners. Social business due to its flexibility and ability to attract material and non-material support resources may offer its services and goods even to the lowest income-receiving persons (bottom of the Pyramid conception).

Social business – business model according to which using the business mechanism pursuit of profit is associated with goals and priorities important for the society, making reference to socially responsible business, partnership provisions of public and private sectors, applying social innovations. Social business comprises three main aspects: enterprise (permanent economic commercial activity), social (ambition for social networks) and management (limited profit distribution) (Socialinio verslo koncepcija, 2015). In a simple way, traditionally, social business refers to private business with a clear social mission.

Social communication – this is universal social, verbal, interactive, social networks' visual or other nonverbal cognition and reflection, with the help of which any social interaction act may take place on every social level and through participants of that interaction it leads to the perception, transmission and dispersion of information.

Social clusters – this is the establishment and implementation mean of social innovations intended for solving crucial social and economic problems of the society. Social clusters are social networks of purposefully formed interested groups which allow to unite policy makers, governmental institutions, non-governmental organizations (NGO), science representatives and private companies or private initiatives, intended to unite the joint efforts for

Smart Cities, Big Data, civic hackers and the quest for a new utopia, W. W Norton & Company: New York the implementation of social innovations and solution of social problems (Kvieskienė, Kvieska, Socialinė klasterystė, 2013, p. 64; Kvieskienė, Kvieska, 2012, p.33).

Social economy – group of private, official companies characterized by the autonomy of decisions and freedom of membership established using the market to satisfy its members needs by producing goods, providing services, insurance and resources when decision-making and distribution of the whole profit or surplus funds to its members, each having a vote, are not directly related to their capital contributions or taxes.

Social economy –the national economy sector establishing the social and economic value seeking to satisfy those social and human needs that are not met by private and public sectors. Order of the Minister of Economy of the Republic of Lithuania on April 3rd, 2015, No.4-207 on the Confirmation of the Social Business Conception.

Social economy and entrepreneur (economy of third sector) – in many world states the term is understood as not private, non-state and non-profit organizations and companies providing services and producing goods that influence well-being of the community and person. (Kvieskienė, Kvieska, 2012);

Social education – refers to fundamentals of humanitarian and social sciences: anthropology, philosophy, economy, geography, history, political science, psychology, sociology, and law. It involves not only those educational subjects who intended for social education of a person but also many other educational subjects with gamification impacts are especially important while stipulating a young persons' social culture and values. Moreover, the social maturity of a child is fostered by the life of the school community: organization of systemic and integrated education process, non-formal education, partnership, communication and cooperation among members of the school community, school relations with local community and development of healthy lifestyle skills[10].

Smart socialization – consists of civic participation, social communication and positive socialization and operates through 3M model (multifunctional, multisectoral and multidimensional paradigm of smart education), which in principle reveals itself through the trend of inclusive, creative and responsible society (Kvieskienė, 2016).

Social innovation – innovative new ideas, products, services, positively changing the behavior and health of the community with the purpose to solve

---

10 Socialinis ugdymas. Internet access: https://www.smm.lt/uploads/documents/svietimas/ugdymo-programos/6_Socialinis-ugdymas.pdf

essential socioeconomic challenges and bring up social capital and responsible civic participation (climate change, resource efficiency, continuous education, health care, healthy aging and decrease of social exclusion) (Kvieskienė, Kvieska, 2013, p.33). Social innovations are new ideas (of products, services and models) intended to satisfy social needs of the society or its members and to build new social relations or initiate cooperation. In other words, all is innovations that aim at better society and stronger ability of the society to act (Murray R., Caulier-Grice J., Mulgan G., 2010).

Social innovations – new behavioral changes that need to deal with the most important socioeconomic challenges, for example, climate change, resource efficiency, security of health care, aging population and reduction of social exclusion (Kvieskienė, Kvieska, 2012, p.33).

Social innovation – this is the development of new social ideas, operating innovative methods and practices, transfer and implementation from other sectors (products, services and mechanisms) in order to satisfy social needs, solve deep-rooted social and environmental problems and establish new social relations and cooperation. The result of social innovations is modern (new), usually nontraditional solutions. The spread of innovations is fostered by the use of information technologies and varied balance of influence on behalf of a single person (consumer). This is the principle applied for building a business or solving the chosen problem.[11] According to description of Phills, Deiglmeier, Miller (2008) social innovation is "a new solution of the social problem which has to be more effective, productive and sustainable compared to the existing solutions, which primarily creates the added value to the whole society and not to the private individuals."

Social life quality – personal well-being and respectful life influencing a person's self-esteem (Giedrė Kvieskienė, at al., 2016).

Social capital – one of society's well-being indicators related to the mutual reliance of different sectors (Giedrė Kvieskienė, Vytautas Kvieska 2012).

Social inclusion – this is a feeling of being part of the community with which you are joined by common identity and values. Public, private and non-governmental organizations' sectoral partnership describes legitimized agreements among state and private sectors developing public infrastructure taking into account the real needs of the community or providing other related services.

---

11 Kvieskienė G., Kvieska V. Socialinės ekonomikos inovacijų įtaka visuomenės gerovei// Socialinis ugdymas, Nr.21 (32). Vilnius: Edukologija, 2012, p. 5 – 16. ISSN 1392–9569.

Socioculture – the society of the nearest space in which the creative process takes place.

Social partnership – this is a formalized cooperation system based on legally binding or informal agreements, cooperated working relations and mutually accepted plans between two or more institutions.

Social partnership (social dialogue) – state, employers and employee's interaction, organizational principle based on cooperation, sharing the responsibility while planning, organizing, implementing and evaluating various professional training activities, measures and processes (Šileikis, 1999; Tissot, 2004; Laužackas, 2005; Petrylaitė, 2006).

Social business (social enterprise) – a kind of business with the main social or environmental purpose and not the goal of increasing the profit for shareholders. Accordingly, the earned profit is reinvested back to business or used for charitable and social activities.

Social business – may involve operational principles and form of the activity. As a process, it manifests itself by the appliance of business principle in order to solve social and environmental problems. As activity form, it is created in order to solve social problems, it reinvests profit instead of paying its partners and at least half of its profit comes from the sale of services or goods. Some social businesses are characterized by greater involvement of employees and equal ownership and decision-making rights division among partners. Social business due to its flexibility and ability to attract material and non-material support resources may offer its services and goods even to the lowest income-receiving persons (bottom of the Pyramid conception).

Social clusters – are networks of interrelated organizations operating on the territorial basis in which at least one organization is the public service provider (Order of the Minister of Economy of the Republic of Lithuania on April 3rd, 2015 No.4-207 on the Confirmation of the Social Business Conception).

Social innovations – innovative new ideas, products, services, positively changing the behavior and health of the community with the purpose to solve essential socioeconomic challenges and bring up social capital and civil society (climate change, resource efficiency, continuous education, health care, healthy aging and decrease of social exclusion) (Kvieskienė, Kvieska, 2013, p.33). Social innovations are new ideas (of products, services and models) intended to satisfy social needs of the society or its members and to build new social relations or initiate cooperation. In other words, all innovations aim at a better society and stronger ability of the society to act (Murray R., Caulier-Grice J., Mulgan G., 2010).

Social clusters – innovative development and implementation measures aimed at social innovation for the topical societal social and economic problems. Social clusters are purposefully formed by interest groups enabling joint efforts of policy makers, government agencies, non-governmental organizations (NGOs), research representatives and private enterprises to effectively implement social innovation and solve social problems (G. Kvieskienė, V. Kvieska, Socialinė klasterystė. 2013, p. 64).

Social stakeholders – an interested party (person or organization) that has agreed to take over part of the social responsibility for the selection of highly qualified human resources and their professional training. Stakeholders are government agencies, educational institutions, employers, learners and teachers (Waters, 2005; Pukelis, 2006, Kvieskienė, Kvieska, 2012).

Simulation – the imitation of the operation of real-world process or system over time.

Simulation – partial recovery of reality that distinguishes only the essential characteristics of real life and transfers a process to a risk-free environment, thereby providing participants the opportunity to learn, gain skills and experience.

Simulations (simulation games) – modern teaching method, during which knowledge, teamwork, communication, leadership and decision-making are being trained and developed.

Smart education – a positive socialization strategy, which relies on positive psychology and positive socialization methodological provision and use of innovative technologies of social partnership approaches family friendly, institutional, community, regional and global levels.

Transparency – the disclosure of the selected goals, their implementing measures and ways of fair and intelligent use of it.

Theory of totally integrated education (TIE) predicts that mental structures formed by learners are expected to be stronger when "knowing that one," "knowing how" and "knowing that" are integrated with learners' emotions and intentions (The theory of totally integrated education (TIE), Theodore W. Frick Professor Emeritus Department of Instructional Systems Technology School of Education, Indiana University Bloomington).

3D (3M) model – multifunctional, multisectoral, multi-criterial social education.[12]
"4P" method aims to meet the needs of consumers via the company's marketing mix, consisting of company-controlled marketing variables: product, price, place and promotion. With respect to "4P" complex elements – variables, the decisions are being made on micro/macro environmental factors. Based on P.D. Pelsmacker, M. Geuens, J.V. Den Berg "Marketing communication."

12 Giedrė Kvieskienė. Sumanioji edukacija-socialinėms industrijoms. Internet access: [PDF] *socialinisugdymas.leu.lt › socialinisugdymas › article › download*

Introduction

Background

The Lithuanian Nation[1] has achieved much in building an independent state. Lithuania has unique characteristics and smart human resources, which should be used to foster competitiveness. This could be achieved only with an agreement on clear and achievable goal to fund human resources and research and innovation (R&I), which would cater to the needs of knowledge-based development. Smart specialization has become a key element of the strategy Europe 2020[2] and Lithuania Progress Strategy 2030,[3] which refers to the guidelines for the next decade in the development of regional innovation systems, also consolidates a "smart-growth principles," "green growth" and "inclusion increase" in the EU's Cohesion Policy. Smart specialization is a strategic and systemic approach for priority education in modern society, which provides together with culture, tradition and religion, and which is a key for sustainable growth and development in the society and community. The creation of economic impact is greatest in traditional creative industries, while adapting skills and business models for the creation of organizations and the value and management of intellectual capital. Different practices of how to separate processes, creativity and economy has evolved independently for many centuries, but in the twenty-first century they acquired a new common denominator – they became a creative economy, with intellectual property as a key element in this new economy character (Levickaite, 2011). Today, organizations are becoming increasingly dependent on creativity, because of the abandonment of the traditional physical raw materials and increasing use of the intangible, intellectual raw materials.

Objective and Scope

The goal of the monograph is created around three key objectives:

---

1 The concept Lithuanian Nation is used in the ethnic, and also in political and civic meanings.

2 Benjamin, D.(ed). Europe 2030. In: https://www.brookings.edu/book/europe-2030/

3 Lithuania's Progress Strategy "Lithuania 2030." In: https://lrv.lt/uploads/main/documents/files/EN_version/Useful_information/lithuania2030.pdf; https://lrv.lt/uploads/main/documents/files/EN_version/Useful_information/lithuania2030.pdf

Objective 1: To develop indicators interfaces between social clustering and social capital development.

Objective 2: To implement for habits positive communication for trust-based clustering in educational Communities.

Objective 3: To choose success stories for standardization discourses, our work is focused on conceptualizations, models, frameworks and case studies. Item being proposed to analyze ISO standardization process, which typically analyzes the case for a focus group analyses to convene and to undertake such work while also addressing issues such as education and public market need.

The monograph identifies areas where evidence or data directly relevant to the evaluation is missing and also indicates where further research is needed, especially when data directly relevant to the evaluation are not available. In this monograph, authors investigate the different ways and indicators in which the objectives of social clustering are being realized, in the local, regional or national level as part of a broader examination of the prospects of social clustering for trust and welfare.

Creativity, trust and social partnership are considered highly important for social innovation and prosperity (Asheim, B., 2006, Huggins, R., Izushi, H., Prokop, D., Thompson, P., 2014). Companies in the social and creative industries, such as universal design, social clustering, social advertising, secondary use of things, software or the gamification, are normally seen as particularly innovative (Lazzeretti, L., 2012, Di Maria, E., Grandinetti, R., Di Bernardo, B., 2012, Kvieskienė, G., Kvieska V., 2012).

This monograph argues the greater use of social economics (SE) and social partnership, including "public-private partnerships" (PPP) and "social communication and mediation" (SCM) that innovatively incorporate all social groups and communities from the private and civic sector for smart socialization and innovative education and fund a smart education for social welfare. The integration of smart education (SE), social communication (SC) strategies and public, private, science and civic partnership (PPSCP) into the education process would help to supplement education-funding sources and reduce budget constraints not only in Lithuania but also in other countries. We have tried to explain the synergy between smart education and innovations, to analyze the impacts of social mediation and conflict management, to analyze smart education on integrated communication and to introduce success stories of smart socialization and social innovations. You can find definitions of social economy, social communication and mediation, smart socialization and social business; links between smart socialization and innovations and strategies and scenarios of smart education. We analyze the impacts of social business to marketing: how to modeling success stories of smart education scenarios and social innovations strategies of social business and social economy. Smart education is positive socialization strategy, which relies on positive psychology (Anthony D. Ong and Manfred H. M. Van Dulmen, 2006), methodological provision and use of innovative technologies of how social partnership approaches family, institutional, community, regional and global levels. In socioecological systems research, interdisciplinary access is always important in order to properly identify the problem and ensure integrative outcomes (Jahn, T., Becker, E., Keil, F., & Schramm. E., 2009). The problems of sustainable development are intrinsically linked to the whole socioecological system that is defined to address them (Figure No 1, page 26). This means that both scientists and entrepreneurs in the research field concerned must be considered as participants in the socioecological system under consideration in terms of innovation in the field. This reality is powerfully evident in Lithuanian regional and community strategies (Neil Adams, Neil Harris, 2005). To assure investment in smart education, social industries, creative economy, social capital and regional and local development theory is related to the social economy (Kvieskiene, G., Kvieska, V., 2012, Kvieskiene, et al, 2016). In this monograph, sustainability is broadly defined to include not only environmental but also educational, social, economic and socioecological issues through social partnerships between the private, public, scientific and civic sectors, initiating community-based social and creative industries. Sustainability in this organization is defined as a social partnership of institutions, organizations and communities to harness natural and institutional resources and networking to implement smart education scenarios.

The **object** of this monograph research is the social clustering as a paradigm of trust and Social Capital, focused on smart and sustainable education.

The goal is to research methods include literature analysis, successful case studies, focus group interviews and survey in experts and responsible civil society members for:

- Firstly, to define smart education and community-oriented tasks in theory, documents, statistics, success stories in the context of the Personality Creativity and Social Innovation.
- Secondly, to analyze the main indicators based on expert interviews about measuring the sustainability and smart educational processes including stakeholders, individual institutions, organizations and regions.

- Finally, to argue that smart education is a central indicator for prosperity and universal sustainable development. This is encouraged by additional focus on simulations and education based on creative design. It encloses indicators associated with the needs of local community, meaningful and creative business of children, youth, women and elderly people.

The following **tasks** are addressed to meet these goals:

1. Analyze social economy and social clustering modeling
2. Discuss community-based social industries for creative economy
3. Investigate smart education concept and indicators
4. Scrutinize social clustering and civic engagement interfaces
5. Discuss balancing the culture of the trust building at the institution, community and regional level.

In the monograph, we focus on conceptualization, models and frameworks to combine research and clustering discourses. By analyzing success cases and modeling performance algorithms, a select group of experts refines and extends success case scenarios. Success models in Lithuania or other countries are sought to substantiate conceptual modeling and refined by interviewing experienced experts. As models of social clustering are in the process of being formed, the monograph attempts to define conceptual domains by identifying constructs.

The research question is what smart education strategies or scenarios can help social clustering and consistent educational leadership by initiating networking and intersectoral empowerment to increase the resources of the education sector.

This investigation will select and reveal 12 smart education indicators reflecting smart educational levels, enabling people to master and operate the 3D/3M (Kvieskiene G., Celiesiene E., 2014) model with multifunctional, multicriterial, multisector approach. All problems we put in a guidance for social economy and social cluster competence model aimed at building trust and pragmatic relations with neighbors, colleagues, fellow country people, especially when we must agree upon the important goals. Trust helps to develop not only family but also community, regional and national prosperity. The successful Scandinavian models are based on trust, transparency, social partnership and social cluster, and networking. Aspiration for common objective, consensus on the most important priorities of the public interest, successful social partnership between the public, private, civic (NGO) sector and scientific organizations mean that we have not learned even the elementary democracy model.

Social economy and social communication models-based research,[4] services and virtual learning modules[5] have been developed by Lithuania University of Education (now: Vytautas Magnus University) researchers together with the social partners since 2009. New cluster concept should be brought in when developing awareness-raising initiatives and programs in different fields and on diverse issues such as, for instance, entrepreneurship and corporate social responsibility, anti-discrimination or partnership. Organizations and actors in the social economy sector, in turn, should be open to cooperation with other traditional enterprises and the public sector.

Within the areas of lifelong learning, social capital and trust,[6] Lithuania is still lagging other countries and does not realize that trust and sharing are important not only because of the need to move to a multisectoral multifunctional design, but also for our well-being and health. We have to admit that we lack multisectoral, multifunctional design services, and the community-based service package is yet at an early stage of development.

Keywords: Clustering, Social Innovation, Sustainable development; Smart education; Positive socialization, Private–public Partnership; Trust, Sustainable Development; Entrepreneurship

4 Giedrė Kvieskienė, Vytautas Kvieska. Socialinės partnerystės įtaka Inovacijoms. Monografija. Internet access: http://talpykla.elaba.lt/elaba-fedora/objects/elaba:4318316/datastreams/MAIN/content

5 Viešojo ir NVO sektorių socialinės partnerystės modelis. Internet access: www.nvovaikamskonfederacija.lt/wp-content/.../03/Partnerystes-modelis-Copy.pdf

6 The Legatum Prosperity Index™ 2016. Internet access: http://www.prosperity.com/

I Modeling Social Economy and Social Clustering[1]

1 Guidelines for Cluster Development, a Handbook for Practitioners. Access through the Internet: http://www.minpo.hr/UserDocsImages/Podr%C5%A1ka%20razvoju%20klastera/4.Smjernice%20za%20razvoj%20klastera.pdf

1.1 Community-Based Social Industries for Creative Economy

Lithuania has smart human resources, which should be used to foster competitiveness. Lithuania's economy has grown faster than most other OECD economies over the past 10 years, unemployment continues to fall and public finances have become stable after a long period of deficits and a rising debt.[1] Lithuania's gross domestic product is €34.95 billion a year. That makes it the largest economy of the three Baltic states but just one-tenth of the size of Poland's economy. Small it may be, but it has been growing fast: Lithuania's economy has almost doubled in size since 2000.[2]

Smart specialization has become a key element of the Lithuania Strategy 2030,[3] which lays down guidelines for the next decade in the development of regional innovation systems, consolidates a "smart-growth principles," "green growth," "inclusion increase" and cohesion policy. Smart specialization is a strategic approach to priorities in education and culture, sustainable social services based on tradition and religion as a key for sustainable growth and development in the society.

In this monograph, you will find information about Lithuanian social capital for smart and sustainable socioeducation networking for sustainable society. Social industries, organizations and companies, such as the ones working in universal design, social clustering, social advertising, secondary use of things, software or the gamification are normally seen as particularly innovative (Lazzeretti, L., 2012, Kvieskienė, G., Kvieska V., 2012). Cities and communities are important for social partnerships and social capital: social and creative

---

1 Giedrė Kvieskienė, Vytautas Kvieska (2019). Social Capital for Smart and Sustainable Development in Lithuania. Internet access: https://www.socialnet.de/international/en/papers/social-capital-for-smart-and-sustainable-development-in-lithuania.html#fn1

2 Seven things to know about Lithuania's economy A crash-course on the Lithuanian economy, which becomes part of the Eurozone on 1 January. Internet access: https://www.politico.eu/article/lithuanias-economy-seven-facts-to-introduce-you/

3 Lithuania. 4.4 Inclusive programs for young people. Internet access: https://eacea.ec.europa.eu/national-policies/en/content/youthwiki/44-inclusive-programmes-young-people-lithuania

industries tend to be urban industries, which take advantage of shared knowledge and of a density of specialized customers, suppliers and workers to create new products (Asheim, B. T., Ebersberger, B., & Herstad, S., 2012). Businesses benefit both from the diversity of urban environments, which may provide a range of stimulation, and from specialization, allowed by urban environments. Furthermore, the recent research states that externalities are related to the city size: larger cities provide greater externalities, making firms in large cities more innovative, more inclusive and friendlier for communities (Duranton and Puga, 2003; Stolarick and Florida, 2006, Targamadze 2016). The chapter argues that a greater use of social economics (SE) and social partnership (SP), including "public-private partnership" (PPP) and "social clustering" (SC) an innovative way for empowerment all social groups and stakeholders in communities, in private, public and civic sectors in terms of positive socialization and social welfare. In short, the use of smart education (SE), social communication and mediation (SCM) strategies and the PPP funded education model to complement (but not replace) other sources of funding education could help resolve some of budgetary capital constraints that have been observed and tangibly hampered educational productivity and performance. We have tried to explain the synergy between smart education and social innovations and analyze impacts of social partnership and social clustering on families and community welfare.

Since 1990, the population of Lithuania has shrunk by 23%, largely due to migration, with nearly 72% of emigrants aged between fifteen and forty-four. This has dramatically changed the county's demographic structure and affected the socioeconomical sector in multiple ways. Although Lithuania's economic growth has been impressive, social inequality is still very high; the risk of poverty is one of the highest among European countries, and life expectancy is comparatively low and strongly dependent on socioeconomic background, the sense of confidence and being a member of the community. Low salaries and poor job satisfaction adversely affect the well-being and contribute to the high emigration level. Proper coordination of the labor market and well-balanced social and health policies can all contribute to the improvement of both well-being and economic growth. Lithuania officially claims its priorities including support for social and small business initiatives to create more and better job opportunities, especially for low-skilled employees, and socially protected jobs for disabled people. Better access to state support, social inclusion and an adequate income level combined with support to job seekers and training programs would further facilitate integration of out-of-work individuals into the national labor market. Promotion of equity and effectiveness and sustainability of health policies are also instrumental to inclusiveness.

A wide-scale labor market, unemployment benefits and the pension reform grounded the New Social Model implemented in 2017, which is expected to reinvigorate inclusive growth and underpin the sustainability of public finances in Lithuania.[4] In 2017, despite complex and frequently adverse conditions, improvements were noted in several dimensions of sustainability, and solid fundamentals were laid down for future improvements in several other sectors. Lithuania's non-governmental organizations (NGO) have frequently expressed their intentions to more actively participate in public policies and government decision-making processes. For the first time, umbrella organizations received dedicated government funding and played an important role in advocating for change. In 2013, the European Council decided to start implementation of the Youth Employment Initiative (YEI) to address the issue of young people not in education, employment or training (NEET). Under this initiative, 20 EU member states, where the unemployment rate among 15–24-year-old people exceeded 25% in 2012, received EUR 6.4 billion for implementation of additional measures to reduce unemployment of young people by the year 2018. Lithuania was among the countries and received over EUR 69 million from YEI funds. Lithuania included the YEI into the Operational Program for the European Union Funds' Investments in 2014–2020, where its implementation was assigned a specific objective to reduce the number of young people aged 15 to 29 not in employment, education or training set in Priority and Promoting Quality Employment and Participation in the Labor Market. The objective comprises two YEI-supported projects aimed to support 35,000 young people aged 15–29 who are not in employment, education or training until 2018.[5]

The importance of European cohesion policy is in designing multisectoral, multifunctional, multi-criteria educational, cultural and religious analysis of smart specialization as research and development (R&D) and innovation development priorities are tailored to cater personal, communal, regional, national sustainable development strategy or scenario, considering local resources, competitive advantage and environmental factors (Carlsberg, 2015, Foray, David,

4 OECD Economic Surveys Lithuania June 2018. Internet access: http://www.oecd.org/eco/surveys/Lithuania-2018-OECD-economic-survey-overview.pdf

5 Lithuania. 4.4 Inclusive programs for young people. Internet access: https://eacea.ec.europa.eu/national-policies/en/content/youthwiki/44-inclusive-programmes-young-people-lithuania

Hall, 2009; Kvieskiene, Kvieska, 2012; Kvieskiene, Celiesiene, 2014). After evaluating modern trends, it can be said that Lithuanian society can create a prosperity life. Success will depend on various factors: state policy and financial instruments, smart education and development strategies, investment and efficient management of public assets, multifunctional urban centers, universal design of residential areas and most importantly sustainable development of community participation and engagement. Sustainable development is an important community mobilization tool that develops positive leadership and community capacity, in which it is necessary to validate the local government and other laws (Kvieskiene G., Bardauskiene D., 2014). We define smart positive socialization as social innovation and creative industries synergy, which is based on a 3D model of education in enabling sustainable social communication, smart education, creative Industries and social capital (PPP: social partnership) (Celesiene E., Kvieskiene G., 2014). The creative industries concept emerged in Australia in the early 1990s, but it was given much wider exposure by policy makers in the United Kingdom in the late 1990s, when the Department for Culture, Media and Sport (DCMS) set up its creative industries unit and task force. In the process, the DCMS moved the understanding of the concept of creativity a long way from its common association with activities having a strong artistic component to any activity producing symbolic products with a heavy reliance on intellectual property and created for a wide audience. Emphasizing the key role of intellectual property has classified creative industries into four broad sub-sectors where the copyright, patents, trademarks and design structure are the final product. Social industries are a diverse set of industries, defined as smart education (project managing; gamification and social advertising); sustainable community-based urbanism; sharing and circular economy, secondary use of things[6]; art and antique markets; designer fashion; video, film, and photography; music and the performing and visual arts; publishing; software, computer games, and electronic publishing; radio and television; craft and design.

Totally integrated education (TIE),[7] outreach, experimental education: learning by doing and problem-solving theories predict that learners will better acquire knowledge and skills, strengthen their social emotional health

6 Antrinis daiktų panaudojimas: Internet access: http://idejos.eu/?p=5263

7 Theodore W. Frick (2017). The Theory of Totally Integrated Education: TIE, Indiana University Bloomington frick@indiana.edu (last revised June 29, 2017) Internet access: https://educology.indiana.edu/Frick/TIEtheory.pdf

and resilience if teachers are not alone in learning from the local community resources, explode the socioecological environment and human resources. These theories are based on the theoretical studies of John Dewey, Davide Kolb, Howard Barrows, Ivan Ijich, Maria Montessori, Gary Marx, Bodo van Calsberg, Elizabeth Steiner and John Patrick. Institutions and organizations in the public, private and civic sectors in communities are important for the education of children, but their resources are episodic and underutilized. Smart education processes, educational innovations, social and creative industries emerge in both urban and rural community organizations when leaders of these innovations emerge: entrepreneurs who initiate, facilitate and coordinate this process. Public schools rarely come back to change established educational traditions, and even more so to choose other educational spaces. However, as the experience of the most innovative schools shows, sometimes even remodeling the school space can become an innovative success story. The goal of the school is for children to understand the challenges of real life, to experience the inspirational, supportive and challenging power of not only family and school but the whole socioecological environment and its inspiring magic and potential for development. Learning has long surpassed class and school. Smart education transcends school walls, teaches children to trust in themselves, develops their resilience to life's challenges and teaches them how to find appropriate examples for improvement. The school needs to involve its social partners in the education process so that children can learn from the best, discover a sincere, respectful and trusting relationship with other people, trust their uniqueness and understand their responsibility for development as the basis for the next phase of life. Cities and communities are important for these relationships: social and creative industries tend to be urban industries, which take advantage of shared knowledge and of a density of specialized customers, suppliers and workers to create new products (Asheim, B. T., Ebersberger, B., & Herstad, S., 2012). Businesses benefit both from the diversity of urban environments, which may provide a range of stimulation and from the specialization, allowed by the urban environments. Furthermore, recent research states that such externalities will be related to city size: larger cities provide greater externalities, making firms in large cities more innovative (Duranton and Puga, 2003; Stolarick and Florida, 2006, Targamadze 2016). Globally, creative industries are estimated to account for more than 7% of the world's gross domestic product (GDP) (World Bank 2003) and are forecast to grow on average by 10 % yearly (PriceWaterhouseCoopers 2003). Already these industries represent a leading sector in the OECD (Organization for Economic Cooperation and Development) economies, showing annual growth rates of 5 to 20 per cent

(EESC 2003). In the United Kingdom, for example, creative industries already generate revenues of over £110 billion and employ 1.3 million people (UK Dept. for Culture, Media and Sports 2003). Several other developed countries, such as Australia, Canada, Ireland, New Zealand and Sweden, have also been successful in exploiting their foothold in these industries and are increasingly seeing them as a gateway to the new information economy (United Nations Conference on Trade and Development Distr., 2004). Going toward a sustainable, community-based development highlights a several key elements reflecting the positive impact of 3D/3M:

- the importance of vision,
- enabled social partnership and civil society,
- the critical importance of leadership,
- maximum inclusion of civil society – considering the rating and selection of actions (projects or groups).

Community businesses in many cases are more innovative than others. Businesses may sort into areas based on their relative competitive advantage and would otherwise not survive. The innovative performance of businesses in rural communities' access knowledge from elsewhere and innovate in alternative ways. For example, research on Lithuania case has suggested that community-based innovation is more important for people's welfare and local knowledge spillovers. While some high profile "innovative communities may appear innovative, whether this is because of local linkages or more prosaic profit and sustainability of those innovations, the importance of communities for innovation may be exaggerated". In 2014–2020 in Lithuania, the structural funds will support an integrated regional development, focused on social inclusion, demographical changes stabilization and citizens' involvement (Kvieskiene G., Bardauskiene D., 2014).

Human services, intellectual and social capital are inarguably essential to the twenty-first-century economy, which is dynamic, knowledge based and increasingly global. Indeed, the United Nations Conference on Trade and Development's Creative Economy Report 2010 asserts: "Adequately nurtured, creativity fuels culture, infuses a human-centered [sic] development and constitutes the key ingredient for job creation, innovation and trade while contributing to social inclusion, cultural diversity and environmental sustainability" (Introducing the first report on the Creative Economy of California, 2014).

Smart education is positive socialization strategy, which relies on positive psychology methodological provision and use of innovative technologies of social partnership approaches family, institutional, community, regional and

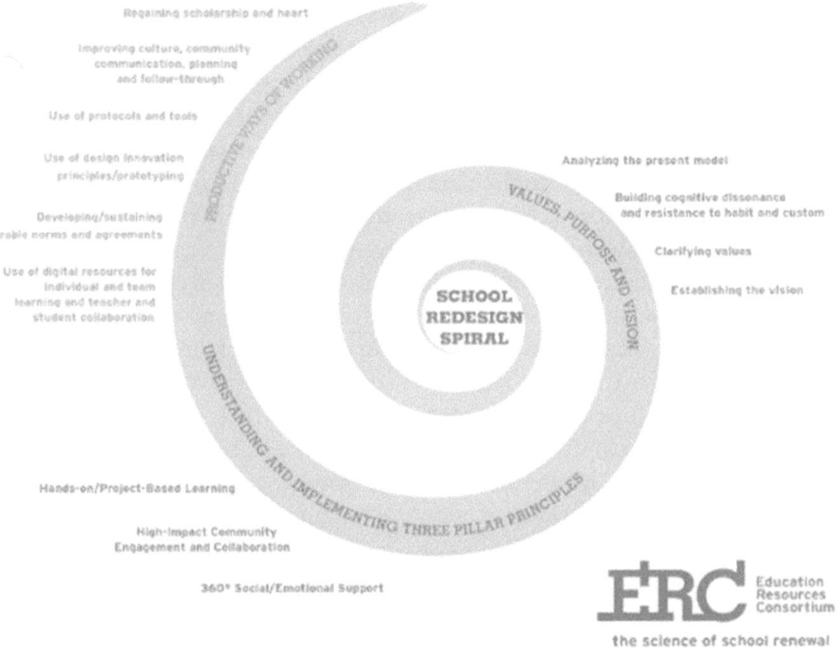

Fig. 1: Inclusive and innovative school model. School Re-Design & New School Innovation. Internet access: http://www.educationresourcesconsortium.org/school-re-design/

global levels. We define smart **positive socialization** as a synergy of social innovation and smart education, which is based on a 3D model of education (multifunctional, multi-criterial, multifactorial) in enabling sustainable social communication, social industries and social capital (Celesiene E., Kvieskiene G., 2014). Over the past decades, youth empowerment models and strategy become increasingly important not only in research but also in the global development organizations studies. The youth policy-oriented research, social networks and international organizations initiative enable young people to participate more in the national and global level decision-making. A systematic analysis of implied and synthesis, which will create new smart education technologies of youth empowerment is based on the theory underlying the adequate formation of young people's competences portfolio of options and measurement criteria. Smart and sustainable education and its sustainability, as well as the quality of education, are widely discussed and controversially perceived in theoretical literature. Smart education as a central indicator of integration and growth is encouraged by additional focus on simulation and education based on creative design. Happy and creative personality is the most important starting point for positive socialization and sustainable education system. Smart education in the practice and theoretical research analyzes peculiarities through quality services, public arena, tradition, fashion, software, tourism, leisure industry and connects it with creativity and social safety. This phenomenon is directly related to creative economy and social capital, which is rapidly increasing through the creation of new jobs, creating new market requirements in response to market expectations. The creative economy and social capital are based on ideas rather than on physical capital; it includes social industries and social communication technologies. The new information, digital content technologies, startups open new spaces and reduce costs. The second feature is increasing interactivity, which bounds together creative view of the creator and the consumer. With iterative master plans, Singapore turned into a hub for international trade, transportation and technology.[8] Adam Smith emphasized the importance of the country's geographic location for the well-being of nations – due to natural conditions, access to transport and geographical location by sea or river – so the economic literature focuses on creation and the distribution of economic wealth. People, groups and countries were unable – as a result of many controllable and uncontrollable factors – to share the available resources – and the income accruing from these resources – in a balanced way.

This has led to unequal or unbalanced economic development, and in many cases, especially from a global perspective, widening the gap between poverty and wealth, not only between people but also between nations. Current concepts and strategies for regional development and regional policy are different from endogenous concepts for regional development in the 1980s and 1990s. At that time, these concepts sought to bridge the gap between urban centers and peripheral areas.

Although this problem remains relevant to Lithuania and other economies in all sectors, especially education and services, alternative approaches, given the development trends that continue to dominate urban centers, appear to be too weak in favor of local innovation strategies. As a result, many rural homesteads or country side tourism regions no longer have the economic strength to

---

8 TUNÇ KARADAĞ. (2013). In Partial Fulfillment of the Requirements for the Degree of Master of Science in City Planning. In: http://citeseerx.ist.psu.edu/viewdoc/download?doi=10.1.1.633.3072&rep=rep1&type=pdf

sustainably develop. Meanwhile, educational innovators, entrepreneurs and civil society see "sustainability" not only as a guiding principle for future development, but also as a prerequisite for systemic access. In order to keep people in hitherto unattractive regions, it is not the infrastructure that needs to be invested in, but innovative ideas that anticipate social partnership, involving and empowering local communities for joint action.

Pragmatic analysis of these regions and successes shows that small businesses or NGOs are leaders in this area. In order to enable schools to do this, it is necessary to adapt systemic funding models, such as the LEADER financial mechanism. In the sustainable development capital stock model (Siegrist et al., 2009 and Brunner et al., 2010), capital includes resources that are located or intended to be created in an economic location for the region to develop in a sustainable way. They represent the intrinsic economic potential of the site; it is important to recognize this potential and develop it in a concerted effort. Capital stocks are four different types of capital that cannot be completely replaced, namely:

1. **Natural capital** (flora and fauna, natural landscape, water, raw materials, etc.) (Lukesch et al., 2010).
2. **Social capital** (trust, values and standards, traditions, engagement and participation, cooperation, networks, etc.).
3. **Economic capital** (real capital in the form of infrastructure, means of production, organizational structures, institutions, business, etc.).
4. **Human capital** (special intellectual knowledge and skills, physical, emotional and mental health, etc.).

Thus, these four capital chairs are the main resource for creating wealth in the region. The challenge of regional development through sustainable development is to identify the potential of the area, raise public awareness and promote participation and exploit capital reserves without destroying them; on the contrary, they must be constantly improved and their synergies sought. Participatory processes support this strategy. To manage this new phenomenon, we analyze scenarios for intellectual positive socialization. Positive socialization (Kvieskienė G., 2005) is understood as an smart educational tool that systematically discovers criteria that ensure the public interest in shaping social and economic policy and decision-making in institutional, political, technological, social, ecological and regional terms.

For managing this new phenomenon, we analyze smart positive socialization scenarios. We understand **positive socialization** (Kvieskiene G., 2005) as a smart educational tool that through systematic institutional, political, technological, social, ecological and regional approach discovers the criteria to ensure the public interest in shaping the socioeconomic policies and decision-making. At the heart of smart education is positive socialization, which develops personality resilience and emotional health, which unleashes the creative forces of the personality and enables independent and responsible functioning and learning. To ensure student positive socialization, for this purpose bringing together integrated team of professionals, inside and outside local community, including outside social partners. Positive socialization is based on the principles of European Commission (EC) White Paper on European Communication Policy (COM (2006) 35, February 2006) defined principles and EC implemented "Plan D" (democracy, dialogue and debate) (COM (2005) 494, 2005 m. October 13) and highlights:

- The principle of openness and transparency. Youth schools must be open and accessible to anyone interested in their activities. This is achieved by ensuring the dissemination of information, accessibility, fairness and competent answers to school community member and social partner questions.
- The principle of inclusion. The members of the school's community (administration, teachers, students and their parents (legal delegates)) and other members of the community or social partners exchange the information necessary for the improvement of student's education (learning) and effectiveness of the school activities. Inclusion aspect also includes dissemination of information in different channels, using information 8 communication technologies, education of community members and social partners, allowing them to acquire skills and abilities to access and use it purposefully. School community members and social partners must be able to express their opinion to be heard, and start the dialogue with decision makers.
- The principle of diversity. It allows schools to establish tolerance to different social and cultural environment perspective.
- Feedback principle. Schools in order to fully and consistently carry out internal and external communication campaigns as well as analyze and evaluate the information submitted by the comprehensibility and (if possible) take into account the views expressed, constantly reflect in organization, in various cases of education (learning), administrating and other actions.[9]

---

9 Alternatyvaus ugdymo institucijose komunikacinis modelis. Internet access: http://www.alternatyvusisugdymas.lt/uploads/2009/12/Alternatyvaus-ugdymo-institucijose-komunikacinis-modelis.pdf

What we analyze in this monograph as a scenario of positive socialization or smart education is related to research on positive psychology and positive socialization that has been gaining ground since 2005, and their links with human neurological predispositions. These chapters devote themselves constantly to neuroscience and applied social science strategies. In neuroscience, social scientists discover an additional opportunity to base sociocultural theory insights on neuroscientific research; in doing so, they seek to develop an interdisciplinary, systematic approach to man. In 1992, Italian neurobiologist Giacomo Rizzolatti discovered mirror neurons, Hopkins University, USA (http://archive.education.jhu.edu/research/nei/news.html), and since 1999, he has been developing and researching neuroeducational programs in collaboration with educators and neuroscientists to find ways to help people change attitude, thinking and behavior. Areas of neuroscience, such as neuroeducation and neurolinguistics programming, are changing psychology and education, affecting social work, nursing and medicine. Educators, social workers and caregivers around the world have been emphasizing since the nineteenth century that empathy, affection and trust-based communication change the behavior of both children and adults. Neuroplastic science has revealed the plasticity of neurons and neural networks in the brain. The twentieth century has proven that even when our childhood ends, neurons that the brain of any age is able to produce new neurons. By 1998, it was thought that we were born with a billion neurons that die slowly with age and that new cells never appear. In fact, millions of unused, unformed stem cells are born and die in our brains. The ability to transform cells into neurons exists throughout our lives, extending the boundaries of this not only in the recovery of post-traumatic health, but also in the correction of behavior in children and adults. Plasticity is observed by scientists not only in the brain, which, using certain methodologies, creates new neural networks and their circuits, but also throughout the human body. Changes occur at the microscopic level. Their appearance is influenced by our daily decisions, and our body only has to adapt to them. This means that every human being has an innate power not only to control his physiology, to change his lifestyle, to adjust his thinking, but also to control the biochemical composition of his body and brain. The question remains as to what methods to use to eliminate or reduce learning disabilities and to adjust adult behavior. When confronted with a child who is unable to learn normally, we focus on helping him or her find individualized ways to help a child with dyslexia, attention deficit disorder, hyperactivity or autism. In the United States, for example, about 8% of children take medication to concentrate and learn. If we find acceptable ways to encourage collaboration between teachers, education professionals, and medics in the event of learning disabilities, we can find acceptable solutions to the problems that arise. In 2017, a unique and only scientific study of this type was conducted in Lithuania, "The Influence of Neuroeducational Techniques on Human Chromosome Telomere Length Dynamics," initiated by the Institute of Personality Development at Raphael Research partners: UAB InMedica ALFA clinic.

The study was conducted at the Life Length Laboratory (Madrid, Spain, http://lifelength.com) by Research consultant professor, doctor of psychology, art therapist Kristine Martinson (Riga Stradins University).

The aim of the study was to evaluate whether the totality of neuroeducational methods developed by M. Mendele-Leliugienė: visualizations, speech codes, tests, exercises and neurodegenerative therapy methods – influence the change of the length of human chromosomes telomeres, November 14, 2017. The results of the research showed that neuroeducational methods developed by M. Mendelė-Leliugienė slow down the human aging processes by 4.2 times. Neuroeducational methods can be applied in practice in psychogenetic as tools of Geno technology. One of them is the belief that each person is best suited to a particular learning style: visual, kinesthetic or contact group learning. Recent research shows that differentiating teaching materials according to different types does not lead to better learning outcomes. The most effective style of learning is supposed to be the pupil's own choice, but this subjective assessment is in many cases based not on style but on convenience. In other words, if the person himself or herself seems to best absorb the information through images, that does not necessarily mean that it is. On the contrary, most information is absorbed through a variety of means and from different environments and subjects.

Over the past decades, citizen empowerment models and strategy become increasingly important not only in research but also in the global development studies. The research oriented to citizen enabling policy, social networks and international organizations initiatives enables people to participate more in decision-making on the national and global levels. A systematic analysis of preferences education and synthesis will create new forms of smart education and of citizen empowerment, which is based on the theory underlying the adequate formation of people competencies portfolio of options and measurement criteria. Smart and sustainable education and its sustainability, as well as the quality of education, are widely discussed and controversially perceived in theoretical publications.

1.2 Smart Education Indicators

Smart Education as a central indicator of integration and growth is encouraged by additional focus on simulation and education based on creative design. Happy and creative personality is the most important starting point for positive socialization and sustainable education system. Creation of economic impact is greatest in traditional creative industries, while adapting skills and business models for the creation of organizations and the value and management of intellectual capital. Different practices of how to separate processes, creativity and economy have evolved independently for many centuries, but in the twenty-first century they acquired a new common denominator – they became a creative economy, with intellectual property as a key element in this new economy character. Today, organizations are becoming increasingly dependent on creativity, because of the abandonment of the traditional physical raw materials and increasing use of the intangible, intellectual raw materials. Creativity is considered highly important for innovation and economic success (Asheim, B., 2006, Huggins, R., Izushi, H., Prokop, D., Thompson, P., 2014). Companies in the creative industries, such as design, publishing, software or the arts, are normally seen as particularly innovative (Lazzeretti, L., 2012, 6. Di Maria, E., Grandinetti, R., Di Bernardo, B., 2012). Cities and communities are important for these relationships: creative industries tend to be urban industries, which take advantage of shared knowledge and of a density of specialized customers, suppliers and workers to create new products (Asheim, B. T., Ebersberger, B., & Herstad, S., 2012). Businesses benefit both from the diversity of urban environments, which may provide a range of stimuli, and from the specialization, allowed by the urban environments. Furthermore, recent research states that such externalities will be related to city size: larger cities provide greater externalities, making firms in large cities more innovative (Duranton and Puga, 2003; Stolarick and Florida, 2006).

It should be emphasized that smart specialization strategies and scenarios prepared in EU financial programming phase (2014–2020) in all the European Union countries. Some of them, like Lithuania, are preparing their own strategy for the entire country, and others are developing separate strategies for different regions or communities. These changes in the socio-political forces help to make a more intelligent analysis of socialization, which calls for a review of existing socialization strategies and practices to ensure compliance with inclusive and creative public education priorities (LSPA feasibility study in 2014) (NLP

Tab. 1: Smart Education Indicators

Sustainability Measurement Index	Acumen Determinants
Education Policy Index (structure: cost of education of (1) the learner in high school; (2) the learner institution of higher education; GDP to science (4), preschool coverage (3), informal (4) education, community/public interest activities (5)	**Identification insight**/intelligence – this is a compromise between each individual's unique skills learning processes and learning individualization in connection with the integration into a common learning space, operating under uncertainty
Education Competitiveness and Social Differentiation Index (structure: the quality of secondary education in scientific production and dissemination (publications, citation index, number of monographs) (1), inclusion rate (students, students with disabilities or special needs) (2), higher education ended a number of people). (PiSSA, PIAAC studies) (3).	**Alternatives and Promotion of Diversity/** Equal Opportunities insight/intelligence – the ability to adapt social integration and inclusion and ensuring the education of socialization.
Integral Sustainable Educational Prerequisites Index: Structure: (1) State welfare ("Legatum" the world's wealth index); (2) Child Welfare (UNICEF index); (3) globalization (Maastricht index); (5) The index of democracy, social networks (clusters) and the number of changes; unemployed and not studying of young people (NEET indicator).	**Impact and Autonomy Insight/** intelligence – this is a problem-solving generation, living in a new or unfamiliar environment in a broader (or interdisciplinary) context of work and academic environments in the context of synergy in education
Smart Modeling of Positive Socialization. Responds to active citizenship in order to prepare the younger generation to evaluate themselves and to project positive social change.	**Positive Socialization Insight/** intelligence – is an integral sustainable educational empowerment through active citizenship initiating positive change.

Association, 2014). In order to integrate the optimal criteria for assessing the sustainability of intelligent education, we have to select 12 of global indexes, which can measure the adaptation of the educational insights of this is crucial for the assumptions and factors on a global, national, regional and individual institutional/organizational levels.

Smart positive socialization as social innovation is based on the idea of the embodiment, based on 3D (multisectoral, multifunctional, multiple criteria) modeling in educational and cultural arenas to simulate sustainable development and social capital. In this way, the smart education running (or) creating breakthrough trends and new opportunities to quickly and creatively use knowledge, innovation, learning, networking and widest possible range of individual and public (community) well-being, with rational costs now and in the future.

Social economy can be influenced by culture and the new challenges arising in respect of the economic crisis and the growing criticism of modernism in relation to social partnership, social innovation, networking, clustering concepts and ecosystemic approach. The ecosystem approach was formulated during the 1990s as a strategy for the integrated management of land, water and living resources (development of the ecosystem approach). The twelve principles were set out at a meeting in Malawi in 1998 (Convention on Biological Diversity, 1998), which was organized under the auspices of the convention on biological diversity (CBD). The social economy is based not only on individual skills, but also on collective knowledge of the various social industries. Successful communities and regions are based on smart, innovative and multifunctional design of environment; public-private partnership, which include social partnership between public, private and civic organizations, a holistic, ecosystem approach. The ecosystem approach is a holistic and inclusive look to the natural environment. It helps to apply current thinking about ecosystem services (what nature does for people) in line with the commonly accepted goal of "sustainable development". It is the primary framework for action under the international Convention on Biological Diversity (CBD).

The ecosystem approach does not replace the many strategies and methods already in place to improve the way people work. Instead, it is intended to inform everyone's thinking about how the environment is connected with society. The ecosystem approach is:

Holistic (considers the environment as a system/an ecosystem);
Inclusive– (recognizes that everyone's knowledge and know-how has a place in decisions); **Adaptive**– (emphasizes that we need to keep on learning, based on experience).

Ecosystem approach encourage education strategists and practitioners to focus more on community oriented, democratic education, which requires the choice of experimental learning through action, experimental education, which requires an innovative solution-based management and democratic education institution-based strategy. These strategies are based on smart education strategies: developed through social partnership, positive socialization, community-based, project- and experimental-oriented methodologies, a bottom-up process involving not only schools but also the local community and other social

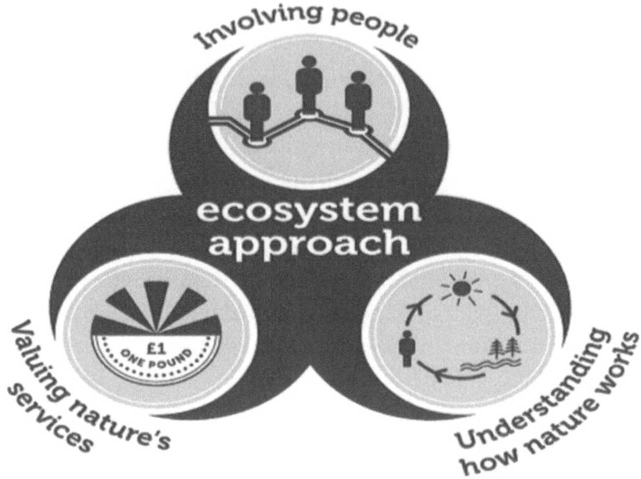

Fig. 2: Ecosystem approach. Ecosystem Approach. In: https://ecosystemsknowledge.net/ecosystem_approach

partners. These educational institutions, not only in Lithuania but also in many other countries, become leaders in certain selected strategies or directions of education, starting with interdisciplinary integration, changing their environment, organization, values, acting on the 3D/3M principle (multisector, multifactor, multifunction) at the community level. This is the principle behind the Citizens' Project[1], which was initiated by the US Civic Education Center. Project Citizen is an active-learning program that engages students in developing public policy for their communities. Project engages teachers in professional development and measures its effects on them and on their students to inspire hundreds of educators worldwide to organize educational activities based on the ideas of democratic education. The Lithuanian College of Democracy (LDUK), a voluntary non-governmental organization, founded in 1997, was seeking to promote the ideas of democracy in society.

The college organizes seminars and develops projects and programs for the promotion of democracy, civic education and development to carry out the program of public education, public debates, publication of material, and promotion of researches. One of those ideas is School of Democratic

1 Project Citizen. Internet access: https://www.civiced.org/programs/project-citizen

Education,[2,3] which Mission are to develop an authentic person who loves himself, others and the world around him and strives for continuous improvement. School is operating in Vilnius for the fifth year, seeks to combine knowledge and action by carrying out common tasks, imitating the educational process, and involving the whole community in decisions. No less inspiring are the schools operating in Vilnius *Gulliver Academy*,[4] which first of all abolished walls inside the school and applied the most modern curricula already developed in the educational process to personalize the educational process. At Gulliver Academy, the educators themselves are involved in the educational process and achieve excellent educational outcomes. "We start our program to develop character by teaching students how to show gratitude," says Rūta Valatkevičiūtė Gustainienė one of the school founders; and from here, we help them understand and adopt values that build on their strengths. Other School founder Stephen Lee is an Australian educator with 35 years of experience in designing, building and leading private schools, and Rūta is a Lithuanian psychologist with 20 years of experience in psychoanalytic approaches to individual and organizational development. Together they have established a unique new school in Vilnius, which is applying modern pedagogical approaches with positive psychology and Tavistockian[5] group work approaches to teaching and learning. The Tavistock Institute of Human Relations UK is a sponsor of the school. Both Stephen and Rūta are graduates of the Institute and are breaking new ground with their initiative to apply group relation theory to the practice in schooling. Happy children are an important goal for both Democracy and Gulliver Academy schools, as we know from research that children learn better when they are happy. It is evidently a more important means than being a part of the academic structure. It is a powerful element in the learning context, because it helps motivate student attitude toward learning in general.

It is recognized in the strategy of smart, sustainable and inclusive growth "Europe 2020" that lifelong learning and development of abilities are the most important elements when responding to current economic crisis and

---

2 Democratic School. Internet access: https://www.demokratinemokykla.lt/

3 Jurgita Adomaitytė Šimavičienė. Mama, noriu į mokyklą. Internet access: https://www.youtube.com/watch?v=QR_JoNj6tlw&list=PLIgCsn0h5WqjBnNPywzeZ4IZM-lfGdEsi&index=2&fbclid=IwAR1JedwY91frkJ-9u8q0Ljp3XV4b4muCSGO0Yo4LFZCvsNSH7-kgl4JuAkc

4 Guliverio akademija. Internet access: http://guliverioakademija.tilda.ws/page3345165.html

5 The Tavistock Institute of Human Relations. Internet access: https://www.tavinstitute.org/who-we-are/

Fig. 3: Sustainable performance based on "working with people". Ecosystem Approach. In: https://ecosystemsknowledge.net/ecosystem_approach

population aging as well as considering a broader economic and social strategy of the European Union.[6] The project is specifically focused on different social power groups in different power groups and sectors that are particularly disintegrated in the social sector which does not have control over responsibility. Therefore, indexes of social exclusion remain unchanged; social sector is divided into separate functions, fulfillment of which is expensive, unified and ineffective. In order for municipalities to legalize the umbrella service sector in a particular geographic area (state, region, municipality and eldership), it is necessary to enable a connecting component (coordination and case manager) who could coordinate public sector, business, NGO and organizations and institutions, which represent them and which operate in a certain field and which are competent to perform social, educational, cultural, health promotion and safety functions and to represent properly a particular settlement, community, region and state. Rudiments of social clusters can be found in the local

6 European Council Resolution on a renewed European agenda for adult learning 2011/C 372/01. Internet access: http://eur-lex.europa.eu/legal-content/LT/TXT/HTML/?uri=CELEX:32011G1220(01)&from=LT

communities or regions where they can be initiated in the best way. They are influenced by the local resources, organizations and institutions performing similar functions and social-ecological infrastructure whose managers should gather together for the common goal. The guidelines have been prepared in accordance with the strengthened successful international and national experience and recommendations.[7],[8] In order for training to meet the requirements of the European documents, we rely on opinion of the European Economic and Social Committee regarding structural and conceptual changes: preconditions are required when developing a globally competitive European industrial structure, which is based on knowledge and research (Europe is passive or leading?)[9] and which specifies the following tasks:

1. To strengthen both cohesion and the cutting-edge results and thus consider the level of economic development;
2. To establish goals and to create measures considering overall nature of many policies (e.g., environment and innovations) and despite this fact to coordinate such policies effectively;
3. To allocate work in the European level and to the member states based on business logic and
4. To establish in the binding manner the measures accepted and to confirm differences.

---

7 See for more details: Guidelines for Cluster Development A Handbook for Practitioners. Internet access:: http://www.minpo.hr/UserDocsImages/Podr%C5%A1ka%20razvoju%20klastera/4.Smjernice%20za%20razvoj%20klastera.pdf; The Unido ApproAch to Business Investment Technology Services Key Principles and Project Experiences for Inclusive Growth Cluster Development. Access through the Internet: https://www.unido.org/fileadmin/user_media_upgrade/What_we_do/Topics/Business__investment_and_technology_services/CUP/UNIDOs_CLUSTER_APPROACH.PDF

8 See for more details: Promotion of sustainable development by strengthening social clustering of public sector: Internet access: http://www.socialcluster.lt/; Partnership of public, private and NGO sector. Research report.2014. Internet access: http://www.esparama.lt/documents/10157/490675/Tyrimas+NVO+akademija.pdf/259b08bc-4236-455c-a315-169b23e4ef42; Model of social partnership of public and NGO sectors. 2014. Internet access: http://www.nvovaikamskonfederacija.lt/wp-content/uploads/2016/03/Partnerystes-modelis-Copy.pdf

9 Communication from the Commission to the Council, the European Parliament, the European Economic and Social Committee and the Committee of the Regions – A lead European market initiative{SEK(2007) 1729} {SEK(2007) 1730}; Internet access: http://eur-lex.europa.eu/legal-content/LT/TXT/?uri=CELEX:52007DC0860

Robert Putnam's books *Making Democracy Work* (1993) and *Bowling Alone* (2000) sparked academic debate among educators, economists, political scientists and sociologists and woke up "renaissance" of Social Policy Studies. Social Economy concept and related activities occurred in Quebec, Canada. The latest socioeconomics, trust civic society trends arose in Europe, North and Latin America, Canada and Africa, today they already exist around the world. The social economy is understood as a pragmatic response to the global economic and social challenges during the crisis. (See: Nancy Neamtan, "Chantier de l'economie sociale" director's report on what is social economy and what are its advantages. *Opinion*. The social economy: link between the market and the state. Canadian example). Scientists Putnam (1993; 2000) and Pierre Bourdieu and James Coleman insights about the concept of social capital (1988) led to the positive sociability, case development studies of personal and public performance, seeking associated power and control, by creation of social networks and communities. High-quality social partnership promotes innovation, and innovation brings unexpected results and initiates new qualitative solutions. Use of the term of socioeconomics in theoretical and social texts that describe a practice is usually associated with social innovations and non-governmental sector (civil society). Phenomena of socioeconomics[10] , social innovations and partnership are discussed in the paper as universal and effective way to combine interests of social partners, to seek a general goal and benefit and to search for new ways and methodologies of problem solving and a new quality of activity. Insights of R. Putnam (1993; 2000), Pierre Bourdieu and James Coleman about conception of social capital (1988) have promoted studies of development of opportunities of positive sociability, personal and public activity when seeking for associative power and will when creating social networks and communities. A quality social partnership promotes innovations and innovations give unexpected results and initiate new quality solutions. Creativity, positive personal social and ecological environment and optimized

10 The term "socioeconomics" and related activities originated in Canadian province of Quebec. New economic tendencies arose in Europe, Northern and Latin America and Africa. Today they exist all around the world. We realize socioeconomics as pragmatic response to global economic and social challenges during a crisis. See: presentation of Nancy Neamtan, the director of "Chantier de l'economie sociale," about socioeconomics and its advantages. Opinion. Socioeconomics: connection between the market and the state. Canadian example: Nancy Neamtan & the Social Economy Movement. Internet access: https://www.youtube.com/watch?v=RzbPoomdHIM

use of human resources increase mutual trust and social capital[11]. In order to implement these challenges, it is necessary to coordinate actions of public, private and non-governmental sectors, to promote a transparent use of resources of public sector, to seek clear priorities and to maximally reduce bureaucratic mechanism. Changes are determined by an accelerating development of science, technique and knowledge that qualitatively changes lifestyle of the modern society. The competitive and learning society can be created by joint efforts of authorities and governmental institutions, political, public and private organizations and all citizens of the country. The society can be analyzed as the market in which people who exchange various things and ideas striving for a general interest and welfare create welfare. Striving to overcome a financial crisis Lithuania did not pay an enough attention to promotion of social involvement and social innovations. Therefore, according to data of Statistics Lithuania, 43% of all income belongs to one fifth of the richest residents of the country and a gap between rich and poor people amounts to seven, 3 times. By this indicator and by a fact that one third of the residents of Lithuania today experience a risk of poverty, we are fourth from the bottom in the European Union. A partial reorganization of the system of social support was started in 2011. This system was not effective enough, and for this reason, initiatives of the citizens to work were insufficient and coverage was increasing insignificantly. A partial solving of problems related to high youth unemployment was started. However, bureaucratic confusion and insufficient transparency could not be avoided in this field too. The expected efficiency was not achieved either. Problems posed by youth unemployment, integration of the groups of social risk, alternative energetics, provision with quality food, health care and aging of the society can be solved only by adopting innovative decisions. One third of the residents of Lithuania today experience a risk of poverty and a long-term exclusion. It is fourth highest rate in the European Union. According to the data of anti-poverty conference[12] and "News week "research[13] (2016), Finland

11 The Dictionary. Human Geography. Ed. R. J. Johnston, D. Gregory & D. M. Smith. Oxford, USA, 1991, p. 559.

12 EAPN. Internet access: https://books.google.lt/books?id=Mq8gDxfR6AMC&pg=PA63&lpg=PA63&dq=Anti-poverty+conference++%E2%80%9Cresearch+(2016)&source=bl&ots=Py8gLyB5Hq&sig=ACfU3U1ag9pHe0in-jXH9eTUP8doaqSx1g&hl=lt&sa=X&ved=2ahUKEwjMzejW9ovqAhWFxIsKHad4DboQ6AEwAnoECAYQAQ#v=onepage&q&f=false

13 Finland trials universal Basic Income on 2,000 Citizens. Internet access: http://www.newsweek.com/finland-trials-universal-basic-income-2000-citizens-538038

best manages to cope with problems of poverty and social inclusion. Why Finland is so well evaluated by two different sources? It is a well-developed country. Its residents receive the highest income. A country is not afraid of challenges. From January 1, 2017, to 2019, 2,000 participants will be paid 560 euros ($587) each month in the hope that it can reduce poverty, cut government red tape and better match societal changes, like increased automation in the workplace. "It's highly interesting to see how it makes people behave. Or, as some critics claim, make them lazier with the knowledge of getting a basic income without doing anything," Olli Kangas, head of the research department at Finnish benefits agency Kela, told the Associated Press. The money is being given to randomly selected unemployed Finns, who will continue to receive the full income regardless of whether they find a job in the next two years. Funding for the experiment comes from the Finnish government, which has pledged 20 billion euros over the next two years. Results will be evaluated in 2019 and other models of basic income may then be explored. Support for basic income came from all of the main political parties in Finland, but critics have said that not enough money is provided to the participants, as the average monthly income of Finnish workers in the private sector is around 3,500 euros. Smaller universal basic income experiments have previously been trailed in the Netherlands, Canada and India, and Glasgow in Scotland is one of several places investigating trial schemes[14]. On the other hand, it is the most educated country and its scholars achieve greatest achievements in mathematics. Finish scientists of education and teachers mostly focus not on intensity of learning but on a safe environment and education background of teachers (all teachers have Master's degree) and individual work (the best and problematic children are taught individually). Every third Finish scholar works individually with a teacher. Singapore can also boast in this field. Although it is one of the smallest states in the world (its area is equal to a half of the territory of New York), it is a country of the most dynamic economy. Good evaluation of Singapore is associated with openness, small taxes and low corruption. The trio of the best countries to live in (only European states) consists of Finland, Switzerland and Sweden. They are followed by Australia, Luxembourg, Norway, Canada, the Netherlands, Japan and Denmark. The United States of America are ranked 11th. Thus, the research analyzing welfare of countries, a level of happiness, a civil index partially confirms a fact that investment in education,

14 Finland trials universal Basic Income on 2,000 Citizens. Internet access: http://www.newsweek.com/finland-trials-universal-basic-income-2000-citizens-538038

socioeconomically innovations and investment in science and its development lead to economic changes and changes of a quality of life. In March 2011, the European Commission started implementation of the initiative of European social innovations whereby it is intended to help developers of innovations in the social field to create new products and services and to develop new methods of work whereby a social value is created and organizations and purchasers are affected. Moreover, this initiative helps find resources and practical knowledge necessary for development of ideas of social novelties into projects or even a new business. Social innovations are one of the measures provided in the communicate "Initiative of social business" dated on October 25, 2011, which was adopted by the Commission. According to this initiative 11 actions to support development of social business and activity in Europe are established, for example, activity related to funding, adaptation of legal systems, marking (hence, also with awareness and visibility) and public procurement. In order to achieve this ambitious goal, public sector and non-governmental organizations have, by using scientific research and technological development (MTTP), promoted demand for innovations. In Europe, many examples of social innovations and innovative ideas can be found in all levels of administration and non-governmental organizations. However, there is a lack of open consultations with social partners that are necessary in order to explore how social innovations and the knowledge economy can penetrate into all level of occupation and all sectors because there is insufficient level of social innovations and social business in the European states. In order to change a situation, analysis of the most successful social innovations and social business was started and a decision was made to strengthen interaction between the program "Horizon 2020" and Cohesion political funds, including additional means in accordance with the program "Horizon 2020" whereby it is intended to bring together more participants, to support development of networks, processes of political thinking and consultations. Innovations that express formation of a new social structure in the society are directly associated with changes of the society's culture and mentality as well with growth of social capital. Changes of a social structure are also closely related to development of social infrastructure: expansion of social infrastructure is a requirement for formation of new social structures and formation of new social structures leads to changes of a social infrastructure and development of social evolution of entire society. Ideas of social welfare for sustainable development are mainly exchanged through social networks using social communication. Therefore, ideas might change, and innovations might occur very fast. Socioeconomics represents global relations of individuals and business by connecting achievements of consumers, business and technological achievements and by invoking a social communication and networks, helps achieve clear qualitative and quantitative results in the daily environment. We analyze behavior socioeconomics using American and European tradition whereby scientists describe socioeconomics as interdisciplinary discipline of social sciences that covers different fields of public life and analyzes their influence on economics and social sector: corruption and anti-corruption policy; consumption, lifestyle, culture and behavior; dark sides of human nature; changes in employment of the labor market; flexibility of the labor market and direct foreign investment; suitability of work, involvement of employees and management; European political integration and progressive economic policy distinguished by American scientists G. Becker, A. Bisin, D. Mortensen and others. In Europe and Canada, socioeconomics is perceived as economically efficient and socially useful activity of the third sector that is in the interim position between public and private sectors. Socioeconomics plays a significant role of power of growth of harmonizing economy and social welfare of the society, considering a fact that almost all fields of welfare today are funded and served by the state. Cooperatives, voluntary associations, communities, charity funds, areas of public and private cooperation declaring the principles of social solidarity are attributed to the field of socioeconomics. In various countries, the boundaries of socioeconomics are perceived in a different way, although many organizations declare use of a profit for harmonization of the society, a democratic way of decision-making and specific fields of work. Analysis of social capital is an important field of socioeconomics. The term "social capital," which was first used in the middle of the last century, defines mutual relations of people, their nature, quality and mutual trust. Social capital stimulates economic development by reducing costs of interaction and at the same time by increasing economic dynamics.

After evaluating modern trends, it can be said that Lithuanian cities can create a full quality of life. Success will depend on various factors: state policy and financial instruments, development strategies, investment and efficient management of public assets, urban centers, residential areas renewal, and most importantly – sustainable development of community participation. Sustainable development is an important community mobilization tool that develops positive leadership and community capacity, in which it is necessary to validate the local government or other laws (Kvieskiene G., Bardauskiene D., 2014). We define smart positive socialization as social innovation and creative industries synergy, which is based on a 3M/D model of education in enabling sustainable integrated communication and social capital (sustainable social partnership) (Celesiene E., Kvieskiene G., 2014).

In youth summer festivals practice, there are excellent examples of positive socialization and smart education in youth culture, which accumulate social capital and social responsibilities. Lithuania became the leader of independent summer youth festivals, which are growing from independent youth group initiatives and based on sharing initiatives, keep natural creativity without impact of power. They are independent initiatives without Government or other pressure and have become excellent examples of social and creative industries, created by youth cooperation, personal impact and activity. For example, "Galapagai," one of the biggest international open-air summer festivals of music and entertainment in Lithuania, inviting music lovers to enjoy live performances and outdoor entertainment in natural environment. Known for its high-quality music, wild atmosphere and a wide selection of entertainment, the festival is growing every year, attracting the participants from Lithuania as well as the guests from Germany, Finland, Sweden, Belarus, Latvia, Poland, Estonia and other countries. This festival takes place at Palūšė, a beautiful corner of nature, where stages are placed to welcome a few dozens of the most famous bands and musicians from all over the world, playing in different styles, so that everyone would enjoy their favorite sound. The festival also contributes to social initiatives and social industries by holding a traditional race of dogs and their owners, as means of encouraging the help to homeless animals.[15] This tradition began in 2007 Lithuania when faced a brand new festival phenomena called **Be2gether**. It was a 3-day lasting festival of music and arts, youth creativity workshops located by the legendary Norviliškės castle. Crowds of young people from all over the country gather there and fall to an absolute relaxation surrounded by nature, nonstop live music and other happy people all spending days and nights in cozy tents or in number of thematic lees while searching for inspiration or simple joy. This festival was the background for hundred initiatives and attracts youth in number of high quality performances by world famous musicians as well as in wide variety of entertainments. The festival participants are being encouraged to free their creativity while attending art classes, to free their madness while bathing in mud, taking an extreme activity like jumping with the bicycle on a hay stacks and to do many other crazy or simply unusual things in order to feel the real independent youth festivals spirit. Be2gether[16] another festival every year created a little town where there is everything what is necessary including wild variety of healthy food cafes and

15 http://galapagai.lt/en/info/about-the-festival/

16 https://www.way2lithuania.com/en/travel-lithuania/be2gether

Picture 1: In YAGA Summer Festival, Lithuania, 2017

bars that are the most popular among youth, night clubs settled in huge tents, a barber, a bathhouse, water entertainments and even the open-air cinema theater. Starting from Be2gether and Yaga (Picture 1), all independent youth festivals propagate nonviolent communication, social responsibilities, healthy food, and the green campaign. The company invites and prompts people to recycle waste, to adapt a green lifestyle and to ride bicycles rather than pollute the earth by driving cars. Those social activities are widely supported by Lithuanian musicians, businessmen and public figures. Regarding the environment, Tundra festival[17] beginning in 2010 is paying big attention to ecology. There were special activities taken by Tundra staffs, which encourage everybody attending the festival to act responsibly regarding pollution while trying to minimize any possible damage for nature during the festival and later on.

Interaction of material, human and social capital ensures a sustainable economic development and determines a coherent development of social innovations. R.D. Putnam was the first to evaluate influence of civil society,

17 Tundra festival: Internet access: https://www.way2lithuania.com/en/travel-lithuania/tundra-festival

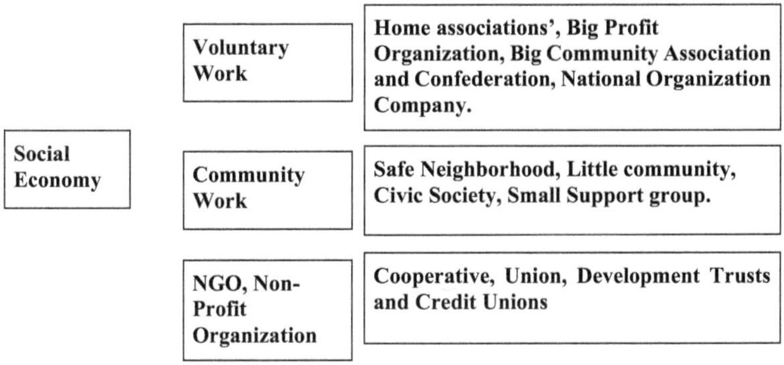

Fig. 4: A structure of socioeconomics. (Kvieskienė, G., Kvieska, V., 2012)

trust and partnership on growth of social capital and associated the latter with the index of civil power. Majority of researchers identify a positive sociability and partnership of the society, authority control and development of social networks as well as development of sociality as the main condition of successful state. Society is understood not only as a fact of a social life but as a value based on trust and a mutual assistance and pursuit of a general goal. It is typical for non-governmental organizations, which, by exploiting a democratic management, achieve excellent results, and best reflect expectations of the citizens.

Socioeconomics (see Figure No 4) promotes development of opportunities of communities and their groups by establishing jobs, by initiating new activities that are necessary for improvement of community and its environment, by providing the locals with opportunities to engage in innovative activities or to take forgotten initiatives. This sector is based on not only economic potential but also the ideas of self-help and cooperation. Socioeconomics involves initiatives of cooperation and self-help; credit unions; household associations; partnerships; community companies and business. Socioeconomics is the fastest growing sector in Europe (see Fig. 4).

Socioeconomics is based on the model of associated economic initiatives and four main values:

- Community-oriented services are more important than the main profit;
- Autonomous management (not controlled by government);
- Democratic decision-making.
- Human and work prevails in respect of a capital.[18]

18 Based on the principles of participation, empowerment. Internet access: http://www.

Empirical measurements of social capital reveal influence of social capital that is directly associated with welfare of the citizens through a social trust, connections of families and the community and a membership in organizations. Religious communities' participation in activity of non-governmental organizations also forms important part of social capital. The society of a lower social capital determines a lower level of economic growth. Social capital reflects important facts of reality, and social networks represent a valuable promoter of welfare and economy. Measurements of conception of welfare often compete with indexes of health, happiness and a social power. Sustainable development requires to refuse from a thought that growth of welfare is only economic expression of welfare. Therefore, discussions regarding a necessity of economic growth and compatibility with welfare of a human and the society reflect two opposite attitudes. Economic welfare, health and happiness have a positive effect of mutual correlation and a causal effect. Research has confirmed that happiness is a reason of good health and affects behavioral and immune system through social relationships, work and other factors. Holistic definition of welfare is one of the most popular indexes of welfare measurement. Creativity, positive person's social and ecological environment and optimized use of human resources increase mutual trust and social capital. To meet these challenges, it is necessary to coordinate the public, private and non-governmental sector actions, to promote transparent use of public funds, pursue clear priorities and minimize bureaucratic mechanism. Development of science, technology and knowledge leads to accelerating changes that in qualitative way modifying modern society lifestyle. Creating a competitive, learning society can be by mutual efforts of authorities and governance institutions, political, public and private organizations, and all citizens of the country. The public can be analyzed as a market in which people create welfare, by exchanging a variety of items and ideas to the general interest and welfare. The most important innovations in social development, which in the experts' point of view are particularly significant in Central and Eastern European societies, are social innovations promoting the emergence of new social institutions and their development; social innovations associated with the development of culture and mentality and social innovations for social services development. Social innovation in the broad sense refers to innovative thinking that sets up citizens' ability to build a future that they crave. Lithuania in order to overcome the financial crisis has not given adequate attention to social inclusion, and to promote social innovation, therefore the richest fifth of the population gets 43% of total income, while the gap between the richest and poorest reaches 7.3 times according to the Department of Statistics. Based on this indicator, and that one-third of Lithuanian residents are now facing the risk of poverty and exclusion in the long term, we are in the fourth place from the bottom in the European Union. In 2011 welfare system is partly reformed, but was not efficient enough, so the citizens' incentives to work were insufficient, while inclusion increased slightly. In part were addressed youth high unemployment problem, but this area was also bureaucratically confused, lacking transparency and lack of expected efficiency. Youth unemployment, social integration of risk groups, alternative energy, provision of quality food and health and aging population posed problems can be solved only by means of innovative solutions. One-third of Lithuanian residents are now facing the risk of poverty and long-term exclusion.

Finland is a developed country that receives the highest income residents. This is the most educated country, and its students reach the greatest achievements in mathematics. Finnish educational researchers and teachers focus not only on learning intensity, but also on safe learning environment and education of teachers (all teachers have a master's degree) and individual work (individual work is done with the best children and children having problems). Every third Finnish student works individually with the teacher.

Singapore has also good achievements. Although it is one of the smallest countries in the world by area – almost half of the New York area – it is the most dynamic economy. Singapore's good ranking is associated with openness, low taxes and low corruption. The best three European countries to live are Finland, Switzerland and Sweden. They are followed by Australia, Luxembourg, Norway, Canada, the Netherlands, Japan and Denmark. United States stays in 11th place. Thus, studies that analyze countries welfare, happiness level and civic index confirm in part that investments in education, social and economic innovation and investment in science and development lead to changes in economic and quality of life. Especially important are social innovations expressing the new social structure formation and their development. The European Commission in March 2011 launched the Social Innovation Europe initiative, which aims to help social innovators to develop new products and services and to develop new working methods that create social value and make impact on organizations and customers. In addition, this initiative will help to find the resources and expertise needed to develop social innovation ideas into projects or even into new business. Social innovation is one of the measures foreseen in October 25, 2011, Communique "Social Business Initiative" adopted by the Commission. Under this initiative 11 actions are foreseen in support of social and business development activities in Europe, that is, related to financing, legal systems adaptation, the labeling (and therefore awareness and visibility) and public

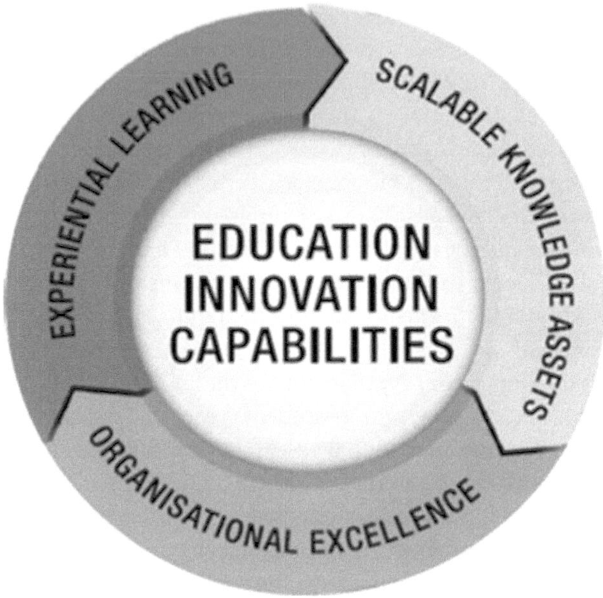

Fig. 5: Education innovation capabilities

procurement. In order to achieve this ambitious goal, the public sector and non-governmental organizations have by using scientific and technological research and development (R & D) to promote innovation demand. In Europe exist many social innovations, innovative ideas and examples at all administrative levels and non-governmental organizations. However, there is lack of open consultations with the social partners to examine how social innovation and the knowledge economy can penetrate into all employment levels and in all sectors, because social innovation and social corporate level in European countries are not sufficient. In order to change the situation EU Commission started analysis of successful social innovation and social business and decided to strengthen the Horizon 2020 and the Cohesion Policy Funds synergies, including additional measures under Horizon 2020 aimed at mobilizing more participants, supporting networking, policy learning processes and consultations.

Innovations that express new social structure formation in society are directly related also to public culture and mentality changes and the growth of social capital. Social structure changes are closely related to the development of social infrastructure: social infrastructure is a prerequisite for the formation of new social structures. Formation of new social structures leads to changes in the social infrastructure and social development of the whole society. Social welfare, sustainable development and change of ideas take place mainly through social networks, taking advantage of social communication, so the change of ideas and the emergence of innovation can take place very fast. The social economy represents individuals and global business ties, includes achievements of consumers, commerce and technologies, and by means of social communication and networking, helps to achieve clear qualitative and quantitative results in everyday environments. We analyze the social economy, taking examples of the United States and the European tradition; some researchers describe the social economy as an interdisciplinary social science discipline, covering different areas of public life, and analyze their impact on the economy and the social sector; corruption and anti-corruption policies; consumption, lifestyle, culture and behavior; the dark elements of human nature; changes in employment, labor market; labor market flexibility and foreign direct investment; the adequacy of work, employees involvement and management; European political integration and progressive economic policies that are described by the United States scientists G. Becker, A. Bisin, D. Mortensen et al. In Europe and Canada the social economy is perceived as the third sector, occupying an intermediate position between the public and private sectors, economically efficient and societal useful activities. The social economy plays a significant role in harmonizing economic growth and public welfare power, taking into account the fact that almost all areas of the welfare today funded by the state. For social economy sector belongs to cooperatives, voluntary associations, communities, charities, etc. In various countries, the social economy limits are understood in different ways, although most organizations declare use of profit for harmonization of society, democratic decision mechanism and specific work field. Important part of social economy is social capital analysis. Use of this concept that began in the middle of the last century of social capital determines peoples' contacts, their nature, quality and trust. Social capital stimulates economic development by reducing interaction costs and thereby increasing the economic dynamics. Materialistic, human and social capital interplay ensures smooth economic development and leads to the development of a coherent social innovation. R. D. Putnam's findings says that civil society requires a certain political environment where individuals are free to discuss and associate in the first place. It likewise shows that trust based socioeconomic development is highly advantageous for all nations attempting to build a strong style democracy. Putnam's says, that at least responsiveness, continuity and sustainability impact of the civil society, role of trust and social partnership on social capital growth, which is directly linked with the civil power index and ability to create welfare state. The main prerequisite for successful state researchers identified the positive public sociability and partnerships, control of authorities, social networking and community development. Community is understood not only as a social fact of life, but also as a value, which is based on trust and self-help, seeking of common objectives. It is characteristic for non-governmental organizations that using democratic governance achieves excellent performance and best reflects the expectations of citizens. Social economy structure (Fig. 4) in the broadest sense consists of various levels of activities that use local communities and groups opportunities for the creation of workplaces, initiating new activities needed for community or to improve the environment, by providing opportunities for local people to undertake innovative activities or initiatives already forgotten. This sector is not only based on economic potential, but also on ideas of mutual help and cooperation. The social economy includes cooperative, mutual help initiatives; credit unions; home associations; partnerships; communities and business enterprises (social economy, volunteering community work, safe neighborhood, small community, civil society, the small support group of housing associations and associations, large business organizations, large community associations and confederations, national organizations and firms, cooperative unions, credit and trust unions, NGOs and non-profit organizations). Social economy – the fastest growing sector in Europe – and the context of diversity and creativity are inexhaustible. The social economy is based on the associated economic incentives model and four core values:

- Service to the community is more important than the main profit;
- Autonomous Management (not controlled by the government);
- Democratic decision-making.
- Human and labor priority over capital.

Social networks and cohesion – this is the objective of modern societies, taking into account people's confidence and desire to jointly develop public welfare. Social capital cumulative index measurements can be conditionally divided into two parts: social cohesion and involvement, as well as community and family networks. In Fig. 4, (57p.) we analyze the variable sub-indices with a breakdown of social capital into several areas. These sub-indices show how the society values volunteering, aid management, donations to charity, organizations impact on the economy and quality of life. Sub-indices also assess the level of public confidence level and the methods by which citizens trust each other, as well as family and religion contributes to the welfare of society. Empirical social capital measurements reveal the influence of social capital,

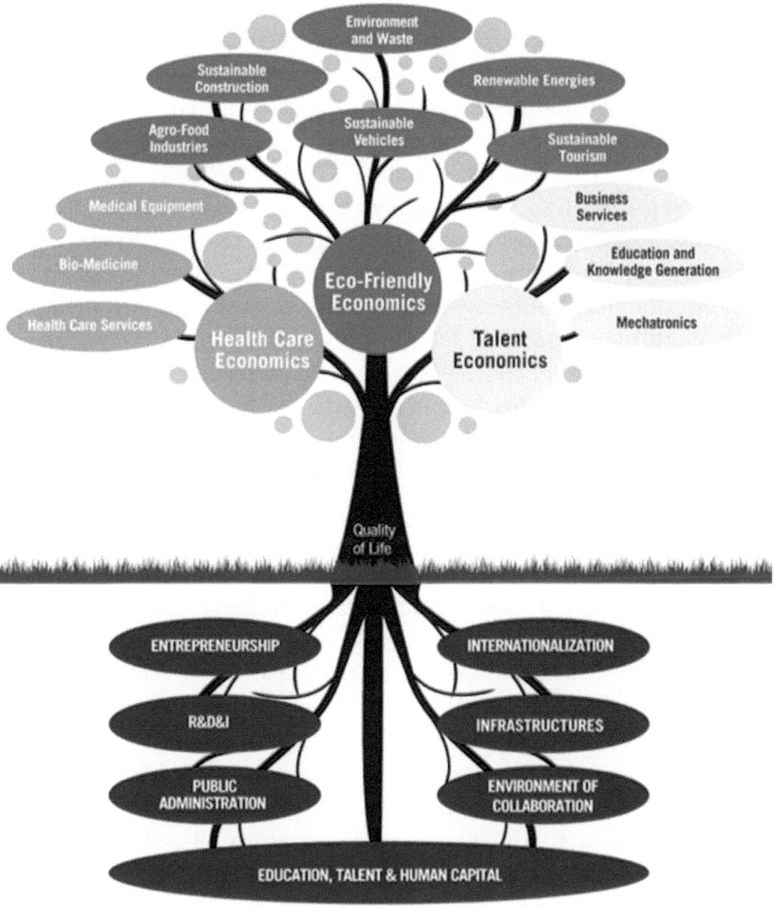

Fig. 6: Smart community development

which is directly related to the well-being of citizens through social trust, family and community, membership in civic organizations. Religious communities' participation in non-governmental organizations is also an important component of social capital. The lower level of social capital in society leads to lower economic growth. Social capital reflects the important facts of reality, and social networks are valuable wealth and economic promoter. Measurements of concept of welfare often compete with health, happiness and civil power indexes. Welfare rate is often measured by assessing the economic level, but Gini coefficient,[19] or World Happiness index[20] argue us, that Nordic countries can be example: the new distribution does not change anyone's health, income, employment, family status, or neighborhood safety… High social capital in Nordic countries societies, than in the rest of European countries (trust based society) confirms to us that trust and ecosystem approaches (Fig. 2) are most important to peoples idea for well-being community and societies. Sustainable performance based on creativity people (Fig. 3), and argue, that sustainable development requires to abandon the thought that prosperity is just an expression of the economic well-being; therefore, the we need economic growth compatibility with the human and social well-being and social capital. Economic well-being, health and happiness have positive correlation and a causal effect. Studies have confirmed that happiness is the reason behind improvement of behavior and the immune system through social relationships, work and other factors. One of the most popular well-being index measurement is a holistic definition of wealth/prosperity. Holistic level of wealth is calculated in terms of economy, entrepreneurship opportunities, taking into account government, education, health, security and reliability, personal freedom and social capital synergies. Definitions of social capital so far vary. Social capital is regarded as an increase of social relationship trust (Fukuyama, 1995, Kvieskienė, 2005), civic involvement (inclusion) created by participation in voluntary organizations (Putnam 1993, 2000, 2015), a public structure that creates a desire to participate in the development of physical capital (Kvieskienė, Kvieska 2012), the aspect of the social structure promoting specific action and forms of cooperation (Coleman, 1988), and the rate of social relations embodied in promoting collaboration between individuals (Fukuyama, 2001). Based on research, the Legatum Institute researchers by evaluating social capital distinguishes 3 classical concepts of social capital groups, which are associated with social cohesion and inclusion (formal volunteering, help to strangers, sacrifice and trust)

---

socialeconomyhub.ca/?q=-content/role-education-promoting-social-economy-canada-%E2%80%93-csehub-thematic-newsletter

19 The Gini coefficient, sometimes called the Gini Index or Gini ratio, is a statistical measure of distribution intended to represent the income or wealth distribution of a nation. The Gini coefficient was developed by Italian statistician Corrado Gini in 1912, and today is the most commonly used measurement of wealth or income inequality. Access by internet: https://worldpopulationreview.com/country-rankings/gini-coefficient-by-country

20 https://happiness-report.s3.amazonaws.com/2020/WHR20.pdf

and civil society (communities, non-governmental organizations and family networks). According to the assessment of social capital, it comprises social cohesion and inclusion (smart community (Fig. 6), formal volunteering, help to strangers, sacrifice, trust and community and family networks, etc.). Lithuania is ranked in 90th place. Rapid economic growth has led to an increase in personal income differentiation, as socially vulnerable groups of the population income growth rate were slower than other population groups. The economic-financial crisis has not reduced, but increased income differentiation growth, because the income for lowest levels decreased faster compared to the highest income level group. Social partnership and innovation (Fig. 5), wider social partnership, inspires creativity, encourages positive personal and social environmental integration, optimizes human resources, and enhances mutual trust and social capital.

In summary, it can be said that the influence of social partnership on development of innovation, social innovation and investment is crucial. Cross-sectoral partnership always initiates new ideas, workplaces, social-eco environment and improvement of quality of life and more qualitative and more efficient social solutions. On the basis of social partnership, social networks are formed, along with other umbrella network organization, enhancing cross-sectoral confidence. Social economic challenge changes the traditional way of thinking and pursues economic and production optimization, development of service economy, together with organized civil society based on social partnership and networking phenomena. High-quality public-private sector partnership improves social-eco environment and people's quality of life, and develops social capital. Participation of civic organizations is improving community life and encourages innovation and investment.

1.3 Trust-Based Social Partnership

Social-economic changes that take place both globally and in Lithuania result in change of skills and competences required in the labor market and social environment. Skills of interpersonal communication and thinking become more important. Skills of effective communication and information exchange are described by the term "social capital." Influence of the human capital on economic growth and innovations was proven long time ago. However, recent macroeconomic research revealed that social capital plays an important role in development of economic and social welfare. By strengthening development of innovations, it can promote economic growth. Moreover, its interfaces with various social aspects, that is, human health, crime rate and number of suicides, were revealed. Trust is necessary for social partnership and network cooperation. It ensures the quality of relationship. Therefore, it is important to communicate not only within the family or at work, but also to take part in social and civic activities. Communication in partnership is the most important human activity. Partnership and trust promote creativity and help to speed up decisions. The more the cooperation, the greater the increase in social capital and in development of social networks.

Data of monograph submitted confirm the fact that compared to other European countries the level of both social capital and social competences is low in Lithuania. This, in turn, affects efficiency of economic performance and leads to negative social consequences. Two main measures of public policy, that is, promotion of public activity and improvement of education quality, increase of its diversity and accessibility, are recommended for improvement of situation. We seek for logic of interventions in pursuance for cooperation between science, business and civil society by empowering already existing new infrastructures (valleys, clusters, incubators, hubs and infrastructures of public institutions); by creating social clusters[1] and by enabling regional and national policies. The

1 See for more details: Cluster policy in Europe. A brief summary of cluster policies in 31 European countries. Internet access: http://www.clusterobservatory.eu/system/modules/com.gridnine.opencms.modules.eco/providers/getpdf.jsp?uid=100146; http://www.europeinnova.org/index.jsp?type=page&lg=en&from=child&classificationId=5969&classificationName=Cluster%20Mapping%20EU15+&cid=5983&parentClassificationId=5967&parentClassificationName=Cluster%20Mapping&parentContentId=5981

system of state aid provision and assessment hitherto poorly promotes cooperation between science, business and civil society. Commercialized products and results generated do not create added value in institutional evaluation; scientific institutions are not promoted to seek synergy of common activities by enriching already existing infrastructure, seeking partnership between science, business, non-governmental organizations and private institutions. One of the essential goals is transformation of the society from industrial to post-industrial knowledge-based society. This is revealed by the study of global future challenges of the EU showing that the existing institutions (governmental, educational, etc.) can no longer adapt too fast and increasingly complex changes of the society[2]. Opportunities of better and more effective response to changes and long-term challenges are increasingly dependent on ability of different social groups to work together in order to achieve common goals.[3] However, this also depends on ability of individuals to adapt, to cooperate with different persons and to work on the basis of appropriate provisions and values. In order to describe a totality of public cooperation capacity, widely recognized authors (F. Fukuyama, R. D. Putnam) have developed the concept of social capital, which rapidly gained popularity worldwide.

The Article 68 of the program of the Government of the Republic of Lithuania:[4]

> The Government poses the following goal in the field of regional policy: to develop an effective system of regional policy, to ensure competitiveness of Lithuanian regions and high quality of life in them. To prevent situation in Lithuania when majority of population is concentrated in one city and to prevent development of regions subject only to agreements entered at the EU level, it is important to strengthen a social, demographic, economic and expert basis in the regions in order each of them would become a strong center of growth and attraction of business and population. Our long-term vision: development of Lithuania with its regions

2 Boden, M. at al (2010). Facing the future: time for the EU to meet global challenges. Joint Research Center Scientific and Technical Reports. Institute for Prospective Technological Studies. European Commission. Luxembourg: Publications Office of the European Union.

3 The policy implications of social capital. Forum report no. 28. National economic and social forum, 2003

4 The Resolution of the Seimas of the Republic of Lithuanian on the program of the Government of the Republic of Lithuania. The 13th of December No. XIII-82. Internet access: https://www.e-tar.lt/portal/lt/legalAct/ed6be240c12511e6bcd2d69186780352

characterized by convenient living conditions, infrastructure that ensures high quality of life, high quality social services, viable economics which takes advantage of the local and international opportunities and creating a dignified life, with the regions which attract investment of the local and foreign businesses, offering a competent labor force and the developed modern infrastructure, with the regions having competent workplaces and, in long-term, regional authorities.

When planning modeling of the social cluster, first of all we will define competences which need to be developed to depict the guidance of preparation of educational measures. Thus, the society's pursuit for changes is limited to vote and expectation for representation. Only every fifth respondent supposes that the government elected can be effectively controlled after elections, for example, by engaging in public political debates, demonstrations or by joining non-governmental organizations and their activity. One third of the respondents recognize importance of personal initiative in the context of social problems solving as effective.[5] When we compare the total indicators in the scoreboard of the innovations union, we can see that we particularly stand behind in the fields of cooperation and enterprise, systems of scientific research and economic impact.

In the wealth index (2018), Lithuania was in the 36 place (Norway 2; Estonia 26; Latvia 37). This place is fairly high. However, (and this is very surprising) even compilers of the index admit that not the real situation but exaggerated pessimism ruins a quality of life in Lithuania. Otherwise, the place could be higher. According to them, Lithuania's economy is growing rapidly, but the residents are still dissatisfied with the country's economy; compared to most other countries, infrastructure of communications are better developed in Lithuania (especially high-speed Internet and use of mobile communications), costs of business establishment are low, but Lithuanians still think that conditions for business are bad; in terms of educational situation Lithuania was in the 43rd place, but in opinion of the locals, the educational system is poor; the healthcare system is adequate, and there are no serious risks in terms of the state security, but anyway Lithuanians feel unsafe in every sense. These conclusions were made in Legatum Institute in London that performs the study.[6]

---

5 Standard Eurobarometer 86 [interactive]. European Commission, 2016 [date of review: the 12th of January 2017]. Internet access: http://ec.europa.eu/COMMFrontOffice/publicopinion/index.cfm/Survey/getSurveyDetail/instruments/STANDARD/surveyKy/2137

6 Welfare (Legatum) index; Internet access: http://www.prosperity.com/rankings

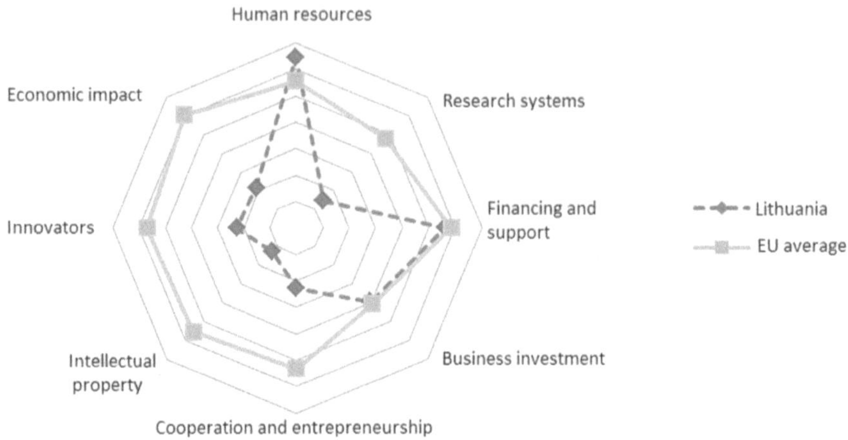

Fig. 7: Cumulative indicators of the Innovation Union Scoreboard: Comparison of the Lithuanian and EU Average.. Republic of Lithuania. Partnership Agreement. Internet access: www.esinvesticijos.lt/uploads/.../files/.../Partnership%20 Agreement_ENG_06%2020.d...

The unstable economic situation, low social capital (trust) level, has a direct impact on families and children living in them. By Prosperity Index (2018) Lithuania ranked in 36th place (Norway 1; Estonia 26, Latvia 40, Germany 14) (Fig. 8). Although the 36th place is high enough, (Fig. 9), it is surprising that even the index compilers recognize that quality life in Lithuania is ruined by "excessive pessimism" rather than the real situation, and the place could/should have been higher.

When we start planning modeling process of the social clustering, first of all we will define competencies that need to be developed to depict the guidelines of preparation of educational measures. The Eurobarometer survey conducted in 2018 shows that half of all young people surveyed suppose that voting in elections is the most effective way of participation in the public life. Smart Education became key strategy of realization as positive socialization strategy, which relies on positive psychology methodological provision and use of innovative technologies of social partnership approaches family, institutional, community, regional and global levels. Positive socialization we understand how smart educational tool that through a systematic institutional, political, technological, social and regional systematic approach in order to discover the criteria to ensure the public interest in shaping the socioeconomic policies and

RANK	COUNTRY									
ADJUST PILLAR WEIGHTING		x1	x1	x1	x1	x1	x1	x1	x1	x1
1	Norway	7	11	3	4	8	1	9	3	8
2	New Zealand	14	2	2	18	17	24	2	1	4
3	Finland	12	6	1	1	25	11	11	14	3
14	Germany	11	12	10	20	24	16	19	16	13
26	Estonia	33	25	20	11	44	38	34	69	6
33	Poland	38	47	38	24	41	23	54	76	47
36	Lithuania	56	58	35	30	72	39	37	98	18
40	Latvia	54	55	46	28	77	41	45	113	16

Econonic Quality
Business Environment
Governance
Education
Health
Safety&Security
Personal Freedom
Social Capital
Natural Environment

Fig. 8: Countries ranked according to Legatum study 9 sub-indices. Adapted from: https://www.prosperity.com/rankings

decision-making. Over the past decades, youth empowerment models and strategy become increasingly important not only in research but also in the global development organizations studies.

When summarizing the need of modeling of competences of social clusters, we can rely on insight of Richard Samans, the board member of the World Economic Forum: "We need to change our way of thinking and appreciate involvement as a positive catalyst of development and wealth. Thus, we need to switch from traditional economic model to incentive of trust, tutelage and integration. A public-private partnership together with scientific studies, endeavor of academic and civil society may help implement the most complex challenges. In pursuance for European welfare development we have to re-evaluate a power of social partnership and trust as hope that this will help us move forward faster." A wide-scale labor market, unemployment benefits and the pension reform grounded the New Social Model implemented in 2017, which is expected to reinvigorate inclusive growth and underpin the sustainability of public finances in Lithuania. In 2017, despite complex and frequently adverse conditions, improvements were noted in several dimensions of sustainability and solid fundamentals were laid down for future improvements in several

Fig. 9: Indexing of Lithuania in accordance with the studies of Legatum Institute. https://www.prosperity.com/rankings

other sectors. Lithuania's non-governmental organizations (NGO) have frequently expressed their intentions to more actively participate in public policies and government decision-making processes. For the first time, umbrella organizations received dedicated government funding and played an important role in advocating for change. In 2013, the European Council decided to start implementation of the YEI to address the issue of young people not in education, employment or training (NEET). Under this initiative, 20 EU member states, where the unemployment rate among 15–24-year-old people exceeded 25% in 2012, received EUR 6.4 billion for implementation of additional measures to reduce unemployment of young people by the year 2018. Lithuania was among the countries and received over EUR 69 million from YEI funds. Lithuania included the YEI into the Operational Program for the European Union Funds' Investments in 2014–2020, where its implementation was assigned a specific objective to reduce the number of young people aged 15 to 29 not in employment, education or training set in Priority of Promoting Quality Employment and Participation in the Labor Market. The objective comprises two YEI-supported projects aimed to support 35,000 young people aged 15–29 who are not in employment, education or training until 2018.

Aspirations for the common objective, consensus on the most important priorities of the public interest and successful social partnership between public, private and civic (NGO) sectors and scientific organizations mean that the country is committed to further development of the introduced democracy model. Education services and virtual learning modules based on studies into socioeconomics, social communication and civic participation have been developed by researchers of Lithuania University of Education and Vytautas Magnus University in cooperation with their social partners since 2009.

Lacking not only the structural units necessary for transformation, but the statehood itself, Lithuania began completely new in 1990. Occupied and annexed by the Soviet Union in the 1940s, Lithuania remained under soviet rule for almost 50 years. The soviet rule dramatically reshaped the country in terms of its resources, economy and, above all, the people. Lithuanians often use the term *rebirth* to describe their return to statehood, which represented the starting point of transformation and set the goal of becoming a free and independent state within the community of European democracies. With the accession to the European Union on May 1, 2004, and the NATO in March 2004, Lithuania finally achieved the goals and aspirations it set in 1990. Sustainable development is an important community mobilization tool that develops positive leadership and community capacity, in which it is necessary to validate the local government and other laws (Kvieskiene G., Bardauskiene D., 2014). We define smart positive socialization as a synergy of social innovation and social industries, which is based on a 3D model of education (multifunctional, multi-criterial and multifactorial) in enabling sustainable social communication, smart education, social industries and social capital (PPP: social partnership) (Celesiene E., Kvieskiene G., 2014).

Tab. 2: Key Indicators

Population (M) 2.9	HDI0.848	GDP p.c., PPP ($)29966
Pop. growth (%p.a.) -1.1	HDI rank of 18837	Gini Index 37.7
Life expectancy (years) 75.1	UN Education Index 0.904	Poverty (%)2.7
Urban population (%) 66.5	Gender inequality0.121	Aid per Capita ($)

Tab. 3: Social Security Rate

Actual	Previous	Highest	Lowest	Dates	Unit	Frequency
41.98	41.98	41.98	33.90	2004– 2018	Percent	Yearly

Lithuania's welfare system is already more than frugal. According (see table 2 and table 3) to the Ministry of Social Security and Labor, only 4% of Lithuania's people receive welfare benefits whereas 9% are unemployed and more than one-fourth live below the at-risk-of-poverty threshold. Lithuania ranks 36th in the overall Prosperity Index rankings. Since the Prosperity Index was introduced in 2007, Lithuania has moved up the ranking table by 11 positions. In the Prosperity Pillar ranking, Lithuania performs best on the Natural Environment and Education pillar and scores lowest on the Social Capital pillar. The biggest positive change over the last year was observed in Social Capital, an increase by 38 positions, while the ranking in Business Environment dropped by 6 positions. In Lithuania, the Social Security Rate (Tab. 3, Fig.10) is a tax related with labor income charged to both companies and employees. Revenues from the Social Security Rate are an important source of income for the government of Lithuania because they help to pay for many social programs including welfare, health-care and many other benefits. Social Security Rate – actual data, historical chart and calendar of releases – was last updated on May of 2019 not all time shows real welfare situation in the Country.

The level of poverty and social exclusion in Lithuania is among the highest in the EU. Statistical data shows that poverty and social exclusion rates in Lithuania are among the highest in the EU (30.1% in 2016, compared to 23.4% EU average).

The Law on Social Enterprises in Lithuania was adopted in 2004 and amended in 2020. Social enterprise is defined by law as any enterprise

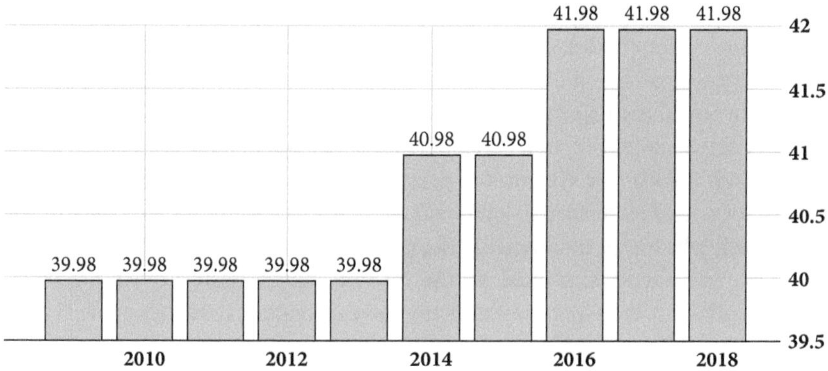

Fig. 10: Social security rate

created to create jobs for vulnerable groups that are disadvantaged in the labor market. The law distinguishes between two types of social enterprise: the social enterprise and the social enterprise for the disabled. Both types of social enterprise must have at least 40% of eligible employees to qualify for social enterprise status and then to qualify for support. It is estimated that 178 social enterprises currently operate in Lithuania, 63 of which are classified as social enterprises for the disabled. They employ over 7.4 thousand people and make up 95% of people with disabilities of the said number. Another 40,000 people with disabilities work in the open labor market, but as many as 120,000 are still in employment – not working at all. The situation is similar in educational institutions where children with difficulties integrate hard. It merely confirms the rule that in order to build a state that is open to all and applies the principles of positive discrimination, we need to cut red tape, provide transparent grants and services to all those in need.

For example, in Lithuania, rapid economic growth has not reduced poverty, has remained stable and the share of people living in low-income households has increased: at-risk-of-poverty rate in 2018 with 22.9% in the country and one of the highest risk of poverty or social exclusion in the EU. The provisions of the Lithuanian Progress Strategy Lithuania 2030 state that it is necessary to develop the creativity of the society and all its members and to emphasize ideas that could help Lithuania to become a modern, dynamic and open state. However, in reality, too much government involvement increases bureaucracy and the welfare system is underfunded. Increasing citizens' inclusion, empowering communities and promoting entrepreneurial innovation are essential to improve the level of social inclusion. This requires schools to be open to special children and to become leaders of social clusters and coordinators of social partners.

Although the welfare system was partly reformed in 2011 to slightly improve the level of social inclusion, its efficiency is still low and the motivation of citizens to seek jobs is still insufficient. The problem of high unemployment among the youth was partly addressed by the reform, but the issues of high bureaucracy, insufficient transparency and the lack of expected efficiency still must be dealt with. The problems of youth unemployment, social integration of risk groups, alternative energy, quality food, health care and aging population can be solved only by means of innovative solutions. One-third of Lithuanian residents are now facing the risk of poverty and long-term exclusion.

The rate of payable unemployment benefits depends on one's employment history. However, unemployment and a low income often overlap. Those on unemployment benefits, especially if they are "low-skilled," can be called to perform public works by local employment offices. Those on income support, if fit to work, are now subject to a requirement to perform socially useful work that benefits the local community. With an underdeveloped unemployment insurance system, social assistance is a key component of the social protection system in Lithuania. After the global financial crisis of 2009, the number of recipients of social assistance increased significantly to 5% of the population in 2014 following the peak of 6.7% in 2012. At the same time, the number of unemployment benefit recipients remained broadly constant, at about 3.5 per cent, and half of the poor are still behind the margins of social assistance. Since 2016, the at-risk-of-poverty rate has significantly increased among old-age pensioners. While the average pay rate has considerably increased, pensions have grown at a much slower pace. As a result, the risk of poverty among old-age pensioners has increased by 6% points. It should also be noted that the average rate of the old-age pension (255 Euro) failed to reach the poverty threshold of 307 Euro set in 2016. Poverty of single parent families and large families also increased significantly.

So far, Lithuania has made some progress in the area of combating poverty, social exclusion and income inequality. However, certain challenges still remain to be dealt with. In 2016, 30% of the Lithuanian population were exposed to the risk of poverty or social exclusion. Therefore, the state started important reforms to increase the rate of retirement pensions, promote employment opportunities, create favorable financial conditions for families and increase the state-subsidized income.

Lithuania's educational system comprises the following stages: 1) early childhood education and care (preprimary and preprimary class-based education); 2) compulsory education for children aged seven through 16 (including primary education, lower-secondary general education, vocational lower-secondary education); 3) upper-secondary and post-secondary education (for persons aged 17 to 19) and 4) higher education provided by universities (undergraduate, graduate and PhD studies) and colleges (undergraduate studies). Lithuania's high level of tertiary attainment has been gradually increasing further in recent years (53.3% in 2014). Its rate of early school leaving is also below the EU average, at just 5.9% in 2014. However, enrollment rates in vocational-education and training programs are low. According to an OECD survey of education released in September 2016, only 15 % of all students are expected to graduate from vocational training programs compared to an OECD average of 46% and EU average of 50 %. Early childhood education attendance is also low, with only 78.3% of Lithuanian children aged four to six attending preprimary education programs, compared to the EU-27 average of 92.3%.

Families with many children, people living in rural areas, the youth and disabled people, unemployed people and elderly people are the demographic groups with the highest poverty risk. A sustainable social network, able to generate more social capital, can only be created through strengthening complex support for families, education institutions, NGOs and schools.

According to the World Giving Index 2016 by Charity Aid Foundation (CAF) Lithuania ranked 124th in the world down from 119th place in 2014. Comparable neighboring countries ranked somewhat higher – Poland – 109th, Latvia – 113th, Estonia – 118th. According to this survey, only 11% of Lithuania's population devoted their time to volunteer activities. From an early age, parents and teachers begin to build trust-based relationships with children that develop their resilience in the face of difficult life situations, empower children to deal with them and seek help from parents or teachers when they need advice or support. Managing social relationships properly is the biggest challenge for every person. It is very important to find support for everyone, especially the child, in the event of failure as they learn the most. The influence of peers increases during adolescence. According to a study by Dario Novak et al. (2018)[7], more than half (55%) of 15-year-olds

7 Dario Novak, a Arunas Emeljanovas, b Brigita Mieziene, b Lovro Štefan, a and Ichiro Kawachi. (2018).How different contexts of social capital are associated with self-rated health among Lithuanian high-school students. Internet access: https://www.ncbi.nlm.nih.gov/pmc/articles/PMC5990934/

trust their classmates, over 80% parents, over 50% school. This study does not explain why similar indicators do not affect the overall social capital ratio, which is among the lowest in many other Eastern European countries (see Fig. 8). Lithuania since 1990 has undergone major political and social changes that have influenced the culture of family life, falling birth rates, rising divorce rates and continuing suicides across all social groups. Education in Lithuania is free and compulsory in primary and lower secondary education (from 6 or 7 years to 16 years). Adolescents may choose upper secondary education at a secondary or vocational school, but may not choose any education after leaving primary school. And while 2018–2019 according to the Lithuanian Social Report, successful implementation of family policy is one of the strategic goals of the Lithuanian government (Ministry of Social Security and Labor, 2015), but children with families are most at risk of poverty and domestic violence does not decrease. The European Parliament calls on the EU Member States to guarantee every citizen the basic right to sufficient resources to have access to social assistance and social services, so that each individual and family maintains his or her dignity and is equally involved in the distribution of social and educational services. Emphasis is placed on accessibility and accountability of social and educational services. Although the role of the Lithuanian Ministries of Social Security and Labor and Education for the well-being of the family cannot be denied, it must be acknowledged that the efforts are non-systematic; projects in the field of family policy do not address the need for essential services and do not provide appropriate positive discrimination support to the weakest: they do not create equal opportunities for all children.

Various national events for families were also organized. Lithuanian's system of training and professional development of social worker's improvement started in 1995. Training centers and institutes of social work have been organizing qualification courses for social workers since 1997. The Ministry of Social Security and Labor launched a program of professional development of social workers in 2001. The main objective of the program is to allow social workers possibilities to model situations in terms of demands, establishment, planning, organizing and evaluation of changes and to develop abilities to search for and effectively apply relevant information. In 2002, the new law on civil services came into force to lay down legal regulation and funding of training and professional development of civil servants and stipulate common principles of social work in view of the contemporary training demands. Professional competences developed by social workers, fill in the gaps observed in professional skills that are characteristic to social pedagogy, such as the absence of teamwork in socioeducational help indicated by G. Kvieskienė, (2003). The analysis states that social work integrates those educational elements that contribute to client's empowerment and, along with psychological and social work elements, they create effective methods of integrated support. At the onset of the third decade of Independence, we can already observe a number of indicators that evidence solid entrenchment of social work in the country, including acknowledgment of social work as a professional practice. The latter has been developing consistently creating and improving its instruments in the historical and cultural context. The main feature of the Lithuanian sociocultural context clearly remains post-communism. In the early stages of development of the welfare state, the post-communist mentality manifested itself in passive labor policies, such as the provision of social services at home, usually to groups inherited from communism, cash-based social assistance and a moderate social service package. For a long time, the quality of social care was measured by technical parameters only (square meters per person, number of bedding sets per year, etc.). Personal satisfaction was not taken into account when emphasizing quality of life (Social Report 2007–2020) and trust in service providers. The European Union countries recognize and respect the right to social security benefits and social services providing protection in cases such as maternity, sickness, accidents at work, dependency or old age and in the event of loss of employment, in accordance with European Union law and national laws and practices. In addition, everyone legally resident and moving within the European Union is entitled to social security benefits and social advantages in accordance with Union law and national laws and practices. Lithuania Declares Citizens' Right to Social Assistance and Housing to Ensure Decent Living in Sufficient Resources in accordance with Union and National Law and Practice to Fight Social Exclusion and Poverty, but in practice these aspirations are difficult to achieve due to the lack of cross-sectoral cooperation between responsible authorities and the low level of networking.[8]

8 Social security and social assistance. Internet access: https://ec.europa.eu/info/aid-development-cooperation-fundamental-rights/your-rights-eu/know-your-rights/solidarity/social-security-and-social-assistance_en

Social work and social Pedagogy – there are life and society changing professions. It is a profession for those, who care about human dignity and social justice. Education of social workers in Lithuania was intensively studied for an entire decade since the first retraining program at Vytautas Magnus University was introduced. The introduction and further development of the new training program caused a lot of discussions. The need to address new challenges in the education and social sectors has prompted Lithuanian educational and social institutions to join global and European networks and initiate international projects that have provided an opportunity to learn from the experience of other Western countries. At the invitation of the Ministries of Social Security and Labor and of Education, Science and Sports and Universities, foreign experts helped develop new programs for teacher training, social work, and specialist development, but there is no agreement on what teachers are needed for future schools. Lithuanian universities, like other Western European universities, focus on postmodern culture in their training of educators or service workers, but do not see the growing significance of networked culture. Initially, the courses were taught mostly by non-professional lecturers specializing in social sciences other than social work and student research projects were done in social fields different from social work. Thus, the social worker's profession was greatly affected by sociology, education, medicine, psychology and other social sciences. On the other hand, social work has been developing both, as a scientific subject and professional practice. Analysis of various sources in the field, including official documents, academic papers, research papers and scientific literature, reveals the presence of contradictory assessments of the pace, scale, quality and authenticity of the evolution of social work in Lithuania. Official evaluations reflected in legislative instruments and primarily based on quantitative factors measured "before" and "after" declaring rapid development of social work as a professional practice. Social work as a discipline entered the qualitative development phase when the net of social work schools built up and social work was included in the classification of major professions. Meanwhile, scientific research comes up with a much smaller and slower progress in the field.

The level of social assistance benefits constitutes another challenge: it has currently become insufficient to alleviate poverty at less than half the poverty line (measured at 60% of the median income). The situation implies an obvious need for improvement in allocating support for out-of-work individuals. A reform of the social assistance system, therefore, appears complementary to the reform of the social model currently being debated in Lithuania. Extending the coverage of the unemployment benefit system, as planned, would provide room for better support to the most vulnerable. The recent debates come up with directions for additional reforms that could complement the implementation of the flexicurity model in Lithuania. A great deal of attention is paid to the timely provision of quality services. Since 2017, the funding for social work with at-risk families has increased to reduce the number of at-risk families per social worker from 17 to 10. The scope of social work with parents of a child in care has been extended to provide conditions for the child's return to the family and increase the number of social workers working with families in local municipalities. Integrated assistance at home (social care and nursing) is granted in all municipalities. Integrated services for parents with children in a crisis have also been improved to grant assistance to parents in the development of their social and positive parenting skills and giving priority to prevention of early family crises. Social clustering and empowered regional policies also help solve social problems in Lithuania. Unfortunately, practical steps required for practical implementation of social partnership and social clustering in Lithuania take much time. Neither adoption of legal acts nor modification of the adopted conception is enough for this. For example, the Law on Concessions of the Republic of Lithuania No. I-1510 adopted on September 10, 1996, is being applied very reluctantly. The laws on and the procedure of complex family assistance and child welfare commissions or councils fail to grant proper intersectional synergy either. The state has developed clear procedures for educational assistance, social support and health care to be provided in a complex way to preschool children and their parents (custodians) and adopted many normative documents regulating the use of resources of European funds and national resources to this end. However, changes in real life are not so impressive since a successful social clustering takes an opportunity of an individual to develop their own competences by strengthening the role of their organizations.

Positive socialization can be described by attempts to negotiate life tasks and set life goals (Weiner, Craighead) by defining an aspect of socialization and, at the same time, promoting democratic values, and as experiences of communication of values helping to establish democratic political institutions (Smidt, Kellstedt, Guth). Giedrė Kvieskienė (2005) defines positive socialization as a policy of positive activities (a scenario of positive socialization) based on which particular persons or their groups adjust behavior of subjects of socialization in order to establish positive components of culture and to prevent consequences of elements of this culture that are unacceptable to the society. Interactive strategies of gamification rapidly gaining popularity in the world should be used when scenarios of positive socialization are applied.

When we try to integrate the optimal criteria for assessing the sustainability of smart education, we decided to select 12 indicators (Tab. 1), which can measure the adaptation of the educational insights of this is crucial for the assumptions and factors on a global, national, regional and individual institutional/organizational levels. In the face of an emerging networking learning ecosystem, many traditional schooling methods are increasingly being called into big question. New finding in neuroscience are reshaping old ideas about performance and cognition. Schools must become the leaders and main actors of design to community based culture, personalized learning experiences, supported by gamification, integrated by subjects and projects, social networking and communication, visualization and active participation in social life activities. Schools must take a central leadership as critical subjects for promoting their communities' socioecological wellbeing. Their leaders must become leaders with new competencies to involve and enabling school and other social partner's communities and learn together innovate, solve problems and to create and transmit knowledge. At the heart of educational institutions over the next ten years will be more and more diversity and difference, and explosion of our current system into a multifunctional, multifactorial, multisectorial or panoramic school model, which prototypes we can find in the best our practices

Vytautas Magnus University after interconnection Lithuanian University of Educational Science became the major teacher training institution in Lithuania. Instead of having an extensive program for future global education of teachers, the University integrates specific topics into the curricula of other programs, for example, a Bachelor's degree program *Social Pedagogics* educating a personality able to find creative ways to respond to current challenges or find the purpose of life under contradictory conditions of the contemporary life. The program integrates such subjects as protection of children rights, civil society and social inclusion, social communication, equal opportunities in education and personal social integration, which serve as the basis for integration of Global Education into teacher training programs. Other Lithuanian higher education institutions include no courses of the global education program for educators into their syllabuses.[9]

9 Augutienė, Indrė, Baltrėnienė, Zina (2014). Global Education in Lithuania Challenges and opportunities. Internet access: https://www.gcedclearinghouse.org/resources/global-education-lithuania-challenges-and-opportunities?language=en

Tab. 4: Smart Education Indicators

Sustainability Measurement Index	Smart Education Determinants
Education Policy Index (structure: cost of education of (1) the learner in high school; (2) the learner institution of higher education; GDP to science (3), preschool coverage (4), informal (5) education, community/public interest activities (6)	**Learning within sustainable environment in changing environment (Identification)**/ Smart education determinants – this is a compromise between each individual's unique skills learning processes and learning individualization in connection with the integration into a common learning space, operating under uncertainty
Education Competitiveness and Social Differentiation Index (structure: the quality of secondary education in scientific production and dissemination (publications, citation index, number of monographs) (1), inclusion rate (students, students with disabilities or special needs) (2), higher education ended a number of people). (PiSSA, PIAAC studies) (3).	**Alternatives and Promotion of Diversity**/ Development of potential for Equal Opportunities insight/intelligence – the ability to adapt socio determinate integration and inclusion and ensuring the education of socialization.
Integral Sustainable Educational Prerequisites Index: Structure: (1) State welfare ("Legatum" the world's wealth index); (2) Child Welfare (UNICEF index); (3) globalization (Maastricht index); (5) The index of Social Capital, Democracy Education, social networks (clusters) and the number of changes; unemployed and not studying of young people (NEET indicator).	**Impact and Autonomy Insight**/Smart Education determinate – this is a problem-solving generation, living in a new or unfamiliar environment in a broader (or interdisciplinary) context of work and academic environments in the context of synergy in education Vocational education and training
Smart Modeling of Positive Socialization. Responds to active citizenship in order to prepare the younger generation to evaluate themselves and to project positive social change.	**Active citizenship and inclusion/Positive Socialization Insight**/Smart Education determinate – is an integral sustainable educational empowerment through active citizenship initiating positive change.

The nature of social work and social education is multifunctional. Thus, social worker's professional practice in health-care institutions is manifold to involve assessment of patient's social situation and needs, evaluation and delivery of social services and rendering advice to the patient and their relatives (advice on social security). The latest studies reveal the importance of social work in health-care institutions, the most important areas of social work and problems most frequently faced by social workers in health-care organizations. Because of the strong interconnection between health and social problems, social workers associate the importance of social work in health-care organizations mainly with the provision of social services aimed at strengthening of individual's capacity of social functioning and granting appropriate social conditions. The key areas of social work include dealing with benefit issues, disability, employment and special needs, informing about social service agencies and organizations, compensation procedures; permanent care and nursing procedures, custody, adaptation of living environment, helping with applications for custody, care or home services, pension matters, mediation between the patient and health-care institutions, liaison with relatives and social organizations, involvement of and social services for family members, organization of employment and self-assistance groups, cooperation with other agencies and professional development.

Lithuanian Social Work association started in 1993. Since then, Lithuania has introduced the national Classification of Professions, adopted necessary legal acts, developed a network of social work institutions, developed a network of training institutions and introduced a clear training system for social workers, launched a system of Professional support, etc. In terms of the narrow meaning, comprising only the elements, social work has already been fully introduced in Lithuania, as A. Bagdonas (2001) states, and there is nothing that can be added to this concern. This periodization of social work in Lithuania suggested by A. Bagdonas (2001) is supported by a vast number of Lithuanian scientists: I. Leliūgienė, E. Giedraitienė, L. Rupšienė (2006), R. Naujaniene (2007), A. Vareikytė (2010), etc. However, foreign scientists who helped to develop social work in Lithuania during the first decade of independence (Ritchie, 2003; Tunney, Kulys, 2004, Lukoseviciene and others) approach the process from a different perspective based on their experience. To them, the question of social work as a profession, raised by A. Flexner in 1915, is still open and unlikely to be answered in the forthcoming decades. The question of professionalization of social work was extremely widely debated in the West from the 60s to the 80s (Greenwood, 1957; Bartlett, 1970, cited in Kavaliauskienė, 2005; Baird, 1972) and the issue is still relevant at the onset of the new millennium, for example, L. C. Johnson (2001) emphasizes the absence of "clear and comprehensive definition" in the context of debates about social work as a profession (cited in Kavaliauskienė, 2005, p. 231). Lithuanian researchers associate emergence and development of the profession with the response to contemporary social needs (Lukoševičienė, 1996; 1999; Bagdonas, 2001; Jurkuvienė, 2003; Naujaniene, 2007; Večkienė, 2007, etc.). The evolution of social work in

Tab. 5: Social services for the Disabled and Elderly people. According to Lithuanian's statistical data of 2015

Recipients of Social Services (Disabled And Elderly People) *	Number	Percentage (%)
Residential care	11159	17,5 %
Social services at home	17826	82,5 %
In day care centers	34500	
In independence-living centers	436	

Lithuania is inseparable from social interventions, associated with the establishment of the welfare state.

In 2017[10], Lithuania had 253.400 persons with disabilities to be supported by the state, 48% of which were men and 52% were women. In 2009, the official number of disabled people registered for the first time started to drop from 27 200 in 2008 to 22 800 in 2009, to 14 300 in 2012, 13 800 in 2013 and 2014, 12 900 in 2015 and 12 000 in 2016.

The most frequent reasons of disability of the working age people included malignant tumors, diseases of the blood circulation system, connective tissue diseases and skeletomuscular system diseases. Mental and behavioral disorders, congenital developmental diseases, deformations, anomalies in chromosomes and diseases of the nervous system are the main causes of children disability. During the period of 2012–2016, the number of persons receiving compensations for attendance (assistance) leveled off (58 300 in 2012, 56 800 in 2013, 56 000 in 2014, 56 400 in 2015 and 56 300 in 2016). On the other hand, the number of the disabled granted compensations for nursing expenses increased during the same period. However, it should be noted that the number has currently become stable: 39 600 compensations in 2012, 39 200 in 2013, 39 900 in 2014, 39 900 in 2015 and 41 300 in 2016. In 2016, compensations for nursing expenses were paid to 942 children. 8 600 children received compensations for attendance (assistance) expenses. The main reasons for children disabilities and special education needs (SEN) include mental disorders, behavioral disorders, hereditary problems, deformations, anomalies of chromosomes and nervous system diseases.

10 Lithuania Data 2016–2017. Internet access: https://www.european-agency.org/data/lithuania/datatable-overview#tab-population_and_enrolment

In 2014, Lithuania adopted a seven-year action plan for transition from institutional care to family and community-based services for people with disabilities and children deprived of parental care. A comprehensive reform of the child protection system is currently being implemented within its framework. Initiated by the Ministry of Social Security and Labor under the slogan *Family for children, community-based services for the disabled*, the reform has a clear goal to dramatically reduce the number of children and people with disabilities receiving institutional care. After the first phase of the reform, which was about both, restructuring and increasing availability and accessibility of services, the reform entered a second phase in mid-2018 focusing on the development of regional infrastructure for family and community-based services.

On April 6–7, 2016, the United Nations Committee on the Rights of Persons with Disabilities discussed the initial report of Lithuania on the implementation of the United Nations Convention on the Rights of Persons with Disabilities in Geneva, and on April 20, 2016, submitted its recommendations to Lithuania. In 2016, financial support was granted to 935 disabled students studying in 37 higher schools including 861 disabled students in 26 public higher schools and 74 disabled students in 11 non-public higher schools. The report by Lithuanian Education and Science states that the total number of students with SEN is increasing in higher education institutions (6,6% in universities and 19, 3% colleges in 2006). A yet, the number of students with special needs declines with each stage of education (15,1% in preschools and only 0,3% at universities). The Law on Education states that "education of learners with special educational needs shall be granted by all schools providing compulsory and universally available education, other education providers and, in certain cases, by schools (classes) established for education of learners with special educational needs." Article 34 of the Law of Education[11] states that "at the request of parents or guardians of a learner with special needs, conditions shall be granted for the learner to study at any preschool education institution or general education school located in vicinity of their home or at any state, municipal or regional school for learners with special educational needs. The pedagogical psychological service shall recommend a school for the child." According to the Law on Education (2011), the purpose of education for learners with special educational needs (SEN) shall be to help them learn and train according to their abilities and attain an education level and acquire a qualification in view

11 Lietuvos Švietimo įstatymo pakeitimo įstatymas. Access in internet: https://e-seimas.lrs.lt/portal/legalAct/lt/TAD/TAIS.395105

of their abilities and capacities. According to the Law, groups of learners with SEN shall be formed according to their special educational needs subdivided into minor, moderate, major and severe. In 2017, learners with SEN in general schools included 66 % minor SEN, 26 % moderate SEN, 6 % major SEN and 2 % severe SEN. Pupils with major and severe SEN may be educated in mainstream general schools (classes) until the age of 21.

Day care services are regulated by the Ministry of Social Security and Labor, but their actual provision is organized at the municipal level. In 2017, there were 222-day care centers in Lithuania, including social care centers, community centers and family support services, which served around 17,300 persons (including children) with various disabilities as well as families at social risk. Providers of day care services are either owned by municipal administrations or run by NGOs. While there is no national level statistics on their geographic distribution, in early 2018, for example, Vilnius municipality had 15-day care services providers for persons (including children) with disability, nine of which were owned by municipal administration and six were run by NGOs (Vilnius, 2018). Municipal providers are financed from municipal budgets, while NGOs can have various funding sources. Generally, municipal administrations purchase day care services from NGOs via public procurement procedures. Specific allocation to NGOs is regulated by a special government decree but it depends on the type of services and their cost (Government of Lithuania, 2006). NGOs can also fund their activities from other sources, including private funds, donations, support from multilateral institutions etc. Day care services are provided by a team of specialists – including social workers, assistant social workers, psychologists and health care, education and employment specialists – at specialized facilities. Services at day care centers are provided 5 days per week and their duration ranges from 3 to 8 hours. The services include informing, consulting, mediating (e.g., with tax or legal authorities), dealing with basic life needs (e.g., preparation of food, hygiene) and transportation as well as more specific ones, such as social and life skills, psychological-psychotherapeutic support, development of working skills (knitting, handcrafts etc.), health-care and other services depending on the needs. The cost of other day care services for individuals is regulated by a government decree and is set up at the municipal level. In general, it depends on the level of income of the service recipient and there are various exemptions for low-income individuals/families.

Statistical data and research in Lithuania reveal that even though the general numbers of inclusive education are growing, the tendencies overall are not showing positive results in the end. Even though the number of children with

SEN in educational institutions is growing, they have a pattern of falling out of the system in their young adulthood, thus resulting in decrease of socialization, communication problems and limited career options.

This tendency points out structural holistic problems in Lithuanian social care and educational systems, opening such flaws as insufficient resources, not adequate inclusion processes and the lack of a good practice model for social care professionals.

Research done by Lithuanian researches and by Lithuanian College of Democracy reveals a considerable user need for more evidence-based recommendations about play for children with disabilities and with special education needs (SEN) and insufficiency of resources to deal with the issue. The first part of this case study discusses the current situation and statistical data relevant to the topic, the second part of the study discusses the improvements and upgrades done in recent years, and the last part of the study describes the rare positive examples of how play is included in daily schedules of children with disabilities and SEN.

Lithuania's convergence process, mainly driven by growth in productivity, has been impressive: GDP per capita rose from one-third to two-thirds of the OECD average level between 1995 and 2015. This was accompanied by a rise in living standards reflecting better job opportunities, education and health (see Fig. 3).

Education in Lithuania is one of the highest in the world. Half of the population has a university degree. Since 1995, Life expectancy has increased by five years. However, despite significant efforts to reduce inequality, it is not declining and the risk of poverty is one of the highest among European countries. Living welfare indicators confirm that in order to move closer to OECD average standards, Lithuania needs special efforts to improve emotional health, labor market flexibility, health prevention, and social and educational equity. High levels of social exclusion and poverty can undermine the sustainability of the convergence process, as shown by recent OECD work on the negative impact of inequality on growth (Causa et al., 2014; OECD, 2015d). Poverty and social exclusion tend to affect the elderly, people with disabilities, children, single-parent households (mainly women) and the unemployed, because the impact of social transfers on poverty reduction is ineffective. Distrust, low social capital is associated with a slow decline in suicide, one of the highest cardiovascular mortality rates in Europe, and the scale of emigration in the last decade. Trust, empathy and a culture of sharing that fosters sharing are slowly changing this situation and increasing social capital. We argue, that social partnership between private and NGO sectors, intensive education and social services

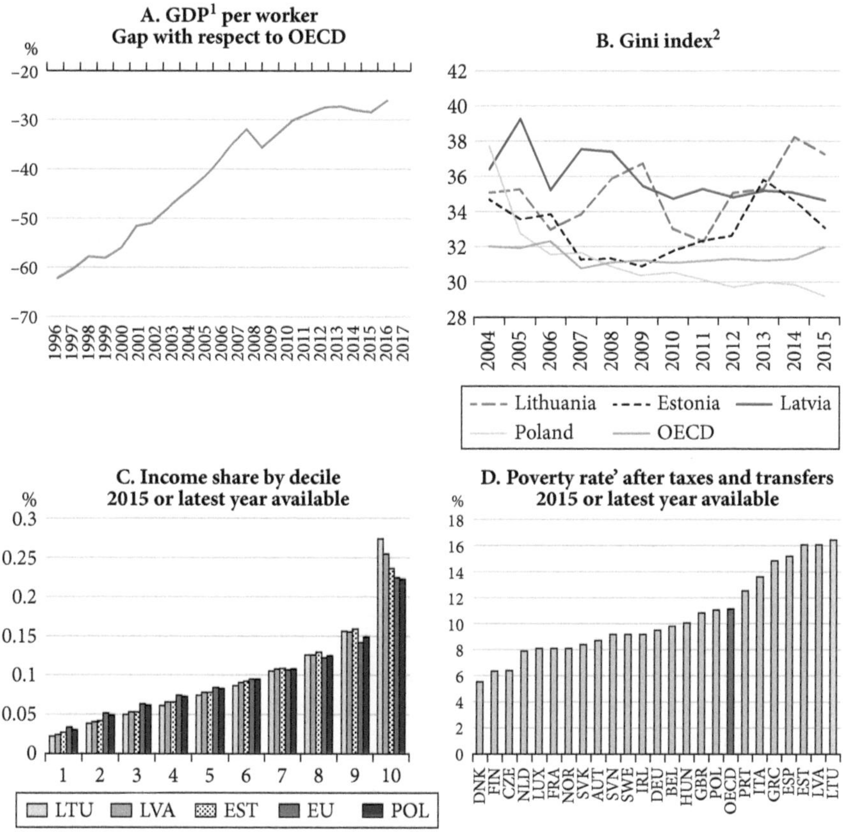

1. GDP in 2010 USD PPP. The rate of convergence is the annual difference in labour productivity gap.
2. Calculated on the basis of disposable income after taxes and transfers.
3. The poverty rate is the ratio of the number of people (in a given age group) whose income falls below the poverty line, set as half the median household income of the total population.

Fig. 11: The convergence process in Lithuania

decentralization and social business integrated in labor market together with social assistance and health policies helps simultaneously strength creativity participation in education and labor market and community well-being. The main messages from the working paper are the following: Good-quality employment is the principal route out of poverty. Further promoting labor market institutions that are conducive to strong job creation, higher employability for the most vulnerable and better job satisfaction are therefore instrumental to promote inclusiveness. Taking measures to further reduce the tax wedge on the low skilled and to promote lifelong learning will all help in this regard.

The risk of unemployment is, however, inherent in a decentralized labor market and is even larger in an economy like Lithuania's, where economic volatility and the convergence process mean labor is frequently displaced. In this monograph, we look for an answer to how cross-sectoral networking promotes the growth of social capital, that is, builds trust between different sectors and broadens active inclusion. The successes we analyze confirm the independence of educational institutions and social service providers, improve the quality of services and expand the social partnership of these institutions. The core idea of our monograph is that the development of social capital through networking, social partnerships, empowering schools to increase the use of socioecological resources, promoting entrepreneurship, improves the quality of education and employment and is the key to poverty. Therefore, we must start with the modernization of educational institutions and the implementation of the lifelong learning model in order to find smart solutions, create new jobs and promote inclusion. The reform of the social model, in line with the principle of flexicurity, must be applied to the educational process. It is very important to start all public sector reforms with the modernization of education, and by balancing the education system with social assistance and health prevention systems, we will be able to seek more and more flexible employment and appropriate means to find work rather than benefits. At present, both in Lithuania and in other countries of the world, these sectors operate independently rather than in cooperation, but in competition with each other, resulting in high spending and a slow decline in poverty and employment growth. International experience shows that better health increases employment, and therefore a healthy lifestyle must be encouraged (beneficial), and unemployment tends to weaken health (Barnay, 2014). Thus, health policy is an integral part of the growth strategy of Lithuania and other countries. Employment protection legislation is strict, but not in line with real life. This increases distrust and undermines job creation. The relatively high proportion of low-wage earners among highly qualified workers also indicates that informal wages ("envelope wages") are common. States do not deal with the unemployment insurance system and provide insufficient financial support for effective job search. It is therefore worth switching to recruitment or job creation through vouchers, scholarships, a basket that all successful small businesses based in regions, non-attractive areas and/or employing socially sensitive individuals (youth, long-term unemployed, returning expatriates, returning from prison, etc.) should be support by EU, national or incubator levels and eliminating bureaucratic monitoring and measuring economic performance. The social support system does not sufficiently reduce the risk of poverty; it is bureaucratic; it does not encourage intersectoral confidence. Sanctions for social assistance recipients are too strict and contribute to the risk of poverty and social exclusion for the most vulnerable. Our life expectancy is highly dependent on geography, socioeconomic conditions. However, the links between health and social capital ratios call for more investment, not in hospital networks, but in health prevention and the development of healthy lifestyles and primary care. Based on the 3D/3M model (multifunctional, multi-criterial and multisectoral), the Positive Socialization Strategy combines active engagement, social entrepreneurship, positive socialization and systematic comprehensive support for the well-being of individuals, families, social groups and communities.

The strategy of positive socialization constructed on the principle of 3D (multifunctional, multi-criteria and multisector) model unites active involvement, social enterprise, positive socialization and systemic comprehensive assistance for welfare of persons, families, social groups and communities. Social capital is a measure of trust, social participation, meaningful networking, culture of sharing and collaboration. It also means open public resources that are easily accessible through social networks and help people achieve their goals, facilitate coordination, communication, problem solving, develop community culture, raise awareness of "we." While social capital pooling usually implies horizontal trust between different groups at the same social level, social capital pooling describes relationships between groups with power differences (e.g., teacher/student/parent/guardian/client, mentor/mentee, administrator/entrepreneur/social worker).

Social capital is a measure of particularized and generalized trust, social participation, integrity and norms of reciprocity. It also represents resources accessed through social networks that help people to achieve their goals, facilitate coordination, communication, resolution of collective dilemmas, reduce the incentives for opportunism, broaden the participants' sense of self, developing perception of "we." The distinction is made between bonding social capital, which results from interactions within a group of people like oneself, and bridging as well as linking social capital, both representing interactions with people from different groups. While bridging social capital refers to horizontal trust among different groups at the same level of social scale, linking social

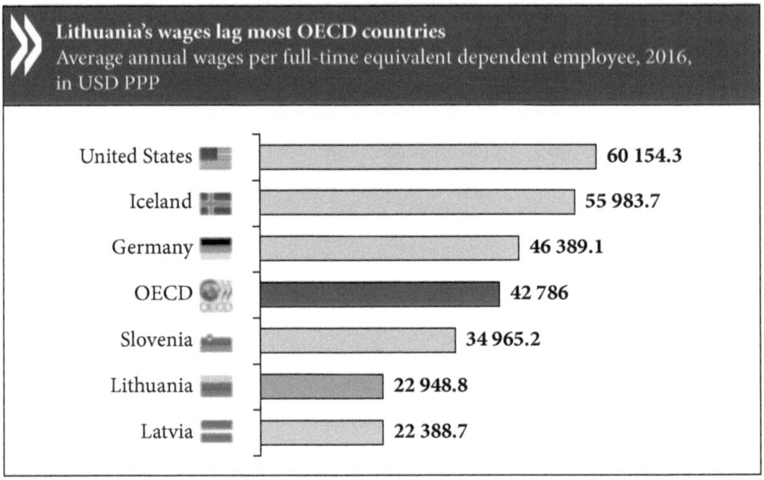

Fig. 12: Lithuania's wage lags most OECD countries

capital describes relationships among groups with power differences (e.g., patron/client or mentor/mentee relationships). This also suggests the multi-contextual nature of social capital, identifying the different social networks that contribute to securing adolescents' sense of belonging. Other studies suggest that, by assessing the relative importance of adolescents' social capital within the family, school, and neighborhood contexts, a better understanding of the impact of these contexts on adolescents' self-rated health will be obtained.

OECD Review of Labor Market and Social Policies: Lithuania says that the population has been declining by more than 1% annually since the early 2000s, as people leave to seek better-paid quality jobs abroad. The working-age population is forecast to decline by about 9% between 2015 and 2020 and by a further 20% in the 2020s. The old-age dependency ratio is thus expected to increase from 28 old-age people for every hundred working-age residents in 2015 to 46 in 2030, a sharper increase than in most OECD countries.

Wage gaps with other European countries are closing too slowly to suggest that emigration will decline soon, according to the report. Income inequality is also very high, and the poorest households have seen little improvement in their living standards over the last years.

"To better share the fruits of economic growth, labor market inequalities and large regional disparities in employment and educational outcomes must be addressed," said Stefano Scarpetta, OECD Director for Employment, Labor and Social Affairs, launching the report in Vilnius with Linas Kukuraitis, Lithuania's Minister for Social Security and Labor. "Moreover, efforts should be made to enhance job quality, given the still widespread informal employment and low wages. The share of informal workers not covered by social security or labor laws is higher than in any other EU country except Latvia." "Becoming a member of the Organization for Economic Co-operation and Development is one of the most important strategic goals of Lithuania," said Linas Kukuraitis, Lithuania's Minister for Social Security and Labor. "We are glad that part of the OECD's recommendations for Lithuania are already being implemented. For example, the reform of Lithuanian Labor Exchange, increasing the adequacy of cash social assistance by increasing the amount of state supported income and introducing the amount of minimum consumption needs calculation methodology." Recent reforms of the Labor Code, the unemployment insurance system, employment policies, minimum-income benefits and public pensions have reduced incentives for informal employment and raised income security. However, resources devoted to labor market and social policies remain limited.

To motivate growing social services and finance active labor market programs, Lithuania can start to active decentralized social market and include more private and NGO sectors to take active part in it. Changing the tax structure to make more freedom for social, family business, community based programs and others social economy tools and balance tax system with society welfare indexes would help for highest quality, as would introducing higher property, capital gains and social environmental welfares. More needs to be done to attract emigrants to come back. Return rates are less than 40% for the migrants who left in 2010, and later emigrants are less likely to return. Among those who do return, more than two-thirds do not work, making them more likely to leave again. The Global Lithuania program, designed to ensure the continued involvement of the diaspora in the life of the state, is currently underfunded and needs to be better resourced, especially to support returning workers. More efforts should also be made to promote job opportunities to Lithuanians unsatisfied with their jobs abroad.

Issues of consumer education have already been integrated into a number of subjects taught in Lithuania's secondary schools; however, the systematic approach toward teaching is still at its onset and needs to be promoted further. Teaching includes little theory and is typically limited to occasionally practical exercises. For example, the syllabus of technology classes includes consumer issues under the topic Culture of Consumption in Estonia, Norway and England and the syllabus of social science includes the topic of consumer rights

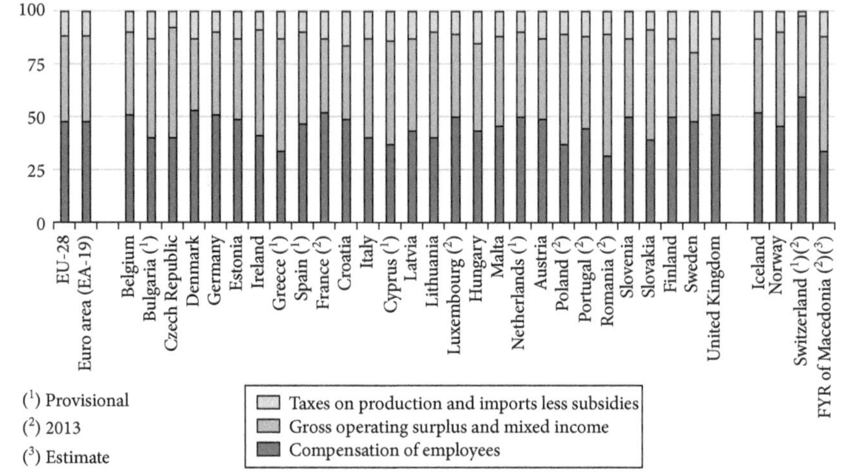

Fig. 13: The relative composition of wages in 2015, the percentage. Source: Eurostat

and responsibilities. On the teacher training level, several graduate courses have been taught, but there is still a need for new attitudes toward the content of teaching and qualifications to be gained. Teaching materials are also few.

EU GDP by income analysis for 28 countries shows that in 2014 the income from production majors from employees' wages, totaling to 47.9% of GDP. Gross operating surplus and mixed income accounted for 40.2% of GDP, while the production and import taxes (less subsidies) was 11.9% (see 11 graphics). Romania (31.3%) was leader in cases of minimum wages of workers to GDP and Greece (33.4%), while 50.0% limit reached or exceeded the seven the EU Member States, which was the highest in Denmark (53.1%).

In the first quarter of 2016 were 122.5 thousand unemployed of 15–74 years of age. During the quarter, the number of unemployed decreased by 7 thousand (5.4%), for the year – 23.3 thousand (16%)[12]. Slow processes affect people's living standards, education, social change and other factors that are related to anticipation of child welfare situation in Lithuania, and they did not allow the implementation of united children welfare measures. Declining unemployment

12 Employment and unemployment in 2016. The first quarter. Internet access: https://osp.stat.gov.lt/informaciniai-pranesimai?articleId=4420181

Tab. 6: GDP in Lithuania[a]

Year	Country's GDP, in bn. EUR	GDP Change, %	GDP/per person, in thous. EUR
2010	28,0	1,6	9,0
2011	31,3	6,0	10,3
2012	33,3	3,8	11,2
2013	35,0	3,5	11,8
2014	36,6	3,0	12,5
2015	37,3	1,7	12,9
2016	38,6	2,3	13,5

[a] Data by the Lithuanian Department Statistics. GDP - current prices. Change - real (adjusted for inflation) growth or decline compared with the previous year, despite of seasonal factors. Updated 2017.

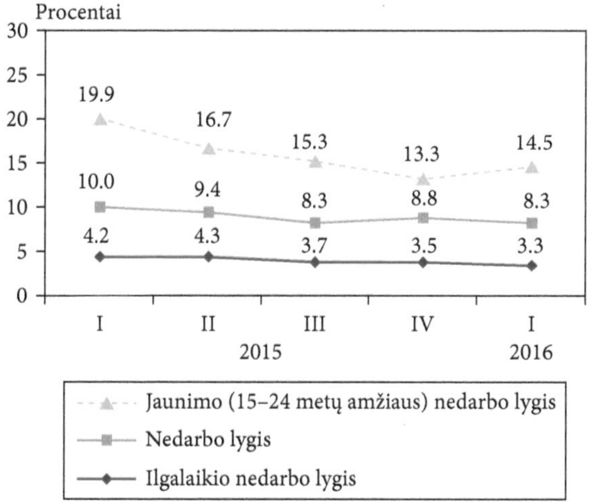

Fig. 14: The unemployment rate in 2015–2016 in quarters. Percentage: Youth (15–24 years) unemployment rate; unemployment rate; long-term unemployment. Employment and unemployment in 2016. The first quarter. Internet access: https://osp.stat.gov.lt/informaciniai-pranesimai?articleId=4420181

is one of the non-governmental and private sector partnerships for child welfare benefit.

As shown by statistics, population and employment is increasing, but for parents who have minor children, are still difficult to find a job, and this affects the number of children at social risk growth. The European Commission on February 20, 2013, issued a communique toward social investment for growth and cohesion – including the implementation of the European Social Fund 2014–2020.[13] Social investments help people prepare for life at risk, instead trying to overcome the consequences of those risks. Social investment plays a special role for people who are disproportionately affected by unemployment, poverty, poor housing and health conditions or discrimination.[14]

One household disposable income in 2015 averaged 845.9 per month and disposable income per household member – 375.9 per month. In comparison with 2014, disposable income increased 7.9%. In the big cities, one household disposable income was 40% higher than in other urban areas and 24.4% higher than in rural areas.[15]

In 2013, young people's (aged 15–24) economic activity was increasing: according to preliminary estimates, one in four of the population was employed, and every fourteenth unemployed. In 2013, 92% was economically inactive population aged 15–24, who were studying in higher, professional institution or secondary school. Youth (aged 15–24) unemployment rate in 2013 fell by 4.8% and equaled to 21.9%. According to preliminary estimates, in 2013 15–24 year-old unemployed were 27.4 thousand – compared to 2012, their number decreased by 4.8 thous. or 14.9%.

One of the largest child welfare issues public, non-governmental and private sector partnerships have to solve is that the schools have too few social pedagogues. According to 2013 data, in 1 193 schools[16] worked 1001 social

---

13 Communication from the Commission to the European Parliament, the Council, the European Economic and Social Committee and the Committee of the Regions. Towards Social Investment for Growth and Cohesion – including implementing the European Social Fund 2014–2020. Brussels, 20.2.2013. 3Ten pat, p. 11.

14 The same, p. 11.

15 Income and living conditions in 2015. Internet access: https://osp.stat.gov.lt/informaciniai-pranesimai?articleId=4662399

16 The number of teachers, pupils and the number of students, the number of educational institutions. Internet access: http://osp.stat.gov.lt/web/guest/statistiniu-rodikliu-analize?portletFormName=visualization&hash=b0740b95-6183-4187-83bb-61da4831ab57

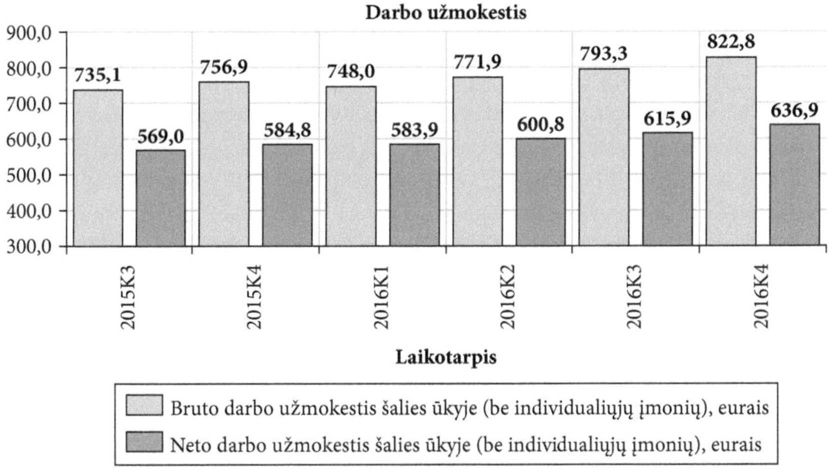

Fig. 15: Remuneration in Lithuania in 2015–2016. Salary, period, gross salary whole in country economy (excluding individual enterprises), in Euros, net salary in country economy (excluding individual enterprises), in euros, Lithuanian Department of Statistics information. Internet access:: http://osp.stat.gov.lt/pranesimai-spaudai?articleId=2478067.

educator. Lithuania teachers' salaries should be calculated by applying rate which differs depending on the experience and training category. Wages increased by 7% for teachers working in early childhood education and care institutions; 5% for young teacher without educational category; an average of 3% for other educators; 2.5% for teaching professionals, such as special education teachers, psychologists and social teachers. Implementing the social partnership is necessary to create conditions for socioeducational assistance in educational institutions that do not have social educators, and socioeducational functions can be taken over the functions of the NGO representatives. According to the Lithuanian Department of Statistics, based on the income and living conditions indicators, estimated risk of poverty in 2012 in the country was 18.6%. Different social groups affected differently by poor revenue. In EU28 in 2014, men and women at risk of poverty (after social transfers) differed quite small: 16.7% men, compared with 17.7% women. In 2014, the greatest men and women of indicators difference was in Estonia (3.2%). According to data provided by Latvia, Cyprus, Lithuania, Sweden and Italy the difference was more than 2.0% too, the same situation was in Switzerland and Norway. Men's risk of poverty rate is slightly higher than for women in seven EU Member States

(Denmark, Hungary, Poland, Romania, Spain, Greece and Slovakia), as well as in Serbia, Iceland and the former Yugoslav Republic of Macedonia. According to income distribution coefficient Lithuania in 2015 was only in 26 place in the EU. Significant differences in income inequality were found between rural and urban regions. The risk of poverty or social exclusion in rural areas is nearly double that of urban areas, which corresponds to the gap in the unemployment rate between cities and rural areas (4.5 % v 11 % in 2017). Lithuania's expenditure on social protection is among the lowest in the EU and the impact of social transfers on poverty reduction is limited, but the recent reforms of social benefits are the first steps in the right direction.[17]

By 2015 data, the economy and the employment rate in the country has improved, but poverty has not declined in the country. The level of people exposed to the risk of poverty or social exclusion in 2015 was as high as 29.3% (2014 m. – 27.3%). This is one of the highest rates among all EU member states[18] (see Tab. 8).

The risk of poverty problem for children (under 18 years) is particularly relevant to Lithuania. Risk of poverty is usually experienced by children up to 18 years. Their risk of poverty in 2015 reached 28.9% and compared to 2014 has increased by 5.4% points. Risk of poverty level for 18–64 year olds compared with the previous year increased by 1.9% points in 2015 and reached 19.5%. During this revenue investigation period (2014), social security costs for families and children increased very little. Increased wages and low unemployment raised income inequality, and children and working-age persons' poverty risk level also increased. Households with children at risk of poverty in 2015 was at 23.3% and compared to 2014 has increased by 3.3% points. Households without

17 COMMISSION STAFF WORKING DOCUMENT Country Report Lithuania 2019 Accompanying the document COMMUNICATION FROM THE COMMISSION TO THE EUROPEAN PARLIAMENT, THE EUROPEAN COUNCIL, THE COUNCIL, THE EUROPEAN CENTRAL BANK AND THE EUROGROUP 2019 European Semester: Assessment of progress on structural reforms, prevention and correction of macroeconomic imbalances, and results of in-depth reviews under Regulation (EU) No 1176/2011. Internet access: https://ec.europa.eu/info/sites/info/files/file_import/2019-european-semester-country-report-lithuania_en.pdf

18 The Government of the Republic of Lithuania, resolution ON GOVERNMENT OF THE REPUBLIC OF LITHUANIA 2016 ACTIVITY REPORT SUBMITION TO THE PARLIAMENT OF THE REPUBLIC OF LITHUANIA. 29 March, 2017. No. 234. Internet access: https://www.e-tar.lt/portal/lt/legalActPrint?documentId=8d5c7320161511e79800e8266c1e5d1b

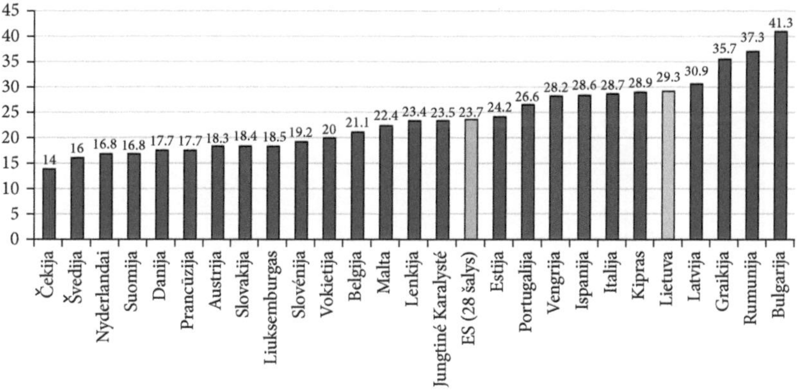

Fig. 16: Individuals who have experienced the risk of poverty or social exclusion in 2015, the percentage. Source: Eurostat

children at risk of poverty in 2015 increased by 3.1% points and reached 21.3%. Looking at the household composition, falling into poverty frequently risked persons living in households that consisted of one adult and dependent children (risk of poverty rate – 47.6%) and single persons (39.5%).[19] A higher poverty level is determined by the following factors: life in rural areas, work in agriculture, poor education, single parenting and number of children in the family. Larger families have a higher risk of poverty than small family. It is obvious that the poor social support for families and children the poverty rate for families with young children, single-parent families and large families, is significantly higher than other families.

The poverty risk group includes families that have children with disabilities or whose parents are disabled themselves. To this group also belong families, where relationships between parents are unstable, parents use alcohol or drugs and not socially motivated to change harmful lifestyles, do not have the skills or desire to fulfill parental responsibilities. In such families, growing skills of children are less developed, and they have fewer opportunities to show their capacities and talents, non-formal education or health services are not available for them, which are necessary for their full development. Therefore, some

19 Poverty risk indicators. Internet access: http://osp.stat.gov.lt/informaciniai-pranesimai?articleId=4601214

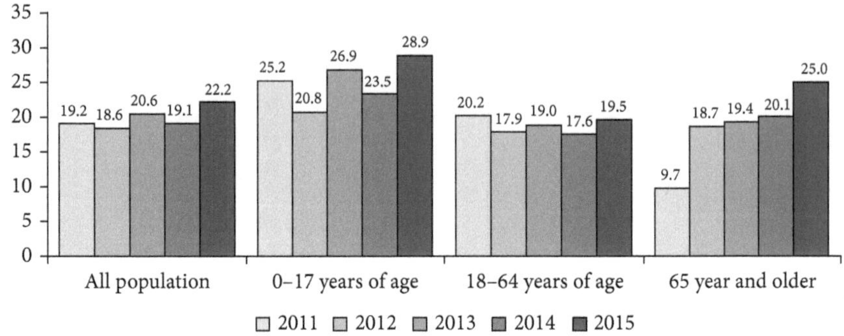

Fig. 17: Poverty rate by age, 2011–2015. All population 0–17 years of age; 18–64 years of age; 65 year and older. Lithuanian Department of Statistics information. Internet access:: http://osp.stat.gov.lt/pranesimai-spaudai?articleId=2478067.

of these children have problems at school: there is no motivation to learn, they have behavioral problems, and etc.

How number of families at social risk has changed in 2001–2015, see Fig. 17.

According to statistics, the dynamics of social risk families is changing slightly. Social investment into families drew attention of many researchers. Social policy, which complements the disadvantaged families' childcare resources, reduces inequality and increase productivity. Many major social and economic problems such as crime, teenage pregnancy, school absenteeism and adverse health conditions are related to the lack of skills and abilities. Skills and abilities education policy can be successful if recognized as multidimension of human abilities. So far, the public policy of cognitive skills usually measured by IQ and other achievement tests. However, these tests ignore important non-cognitive factors that lead to success in learning and life. Cognitive abilities are important for social and economic success. There are socioeconomic, mental health, perseverance, attention, motivation and self-confidence skills. They contribute to the general conduct of society, and even contribute to cognitive measurement results. There are differences in skills between successful ones and the disadvantaged ones in early childhood. Children's family environment (conditions) to a large extent determines further cognitive and socio-emotional abilities, as well as a variety of consequences, such as the future's health and crime.[20] Conditions for families

20 The Case for Investing in Disadvantaged Young Children. By James J. Heckman, Ph.D. (2008), p. 10; James J. Heckman, Ph.D. The Case for Investing in Disadvantaged

in many countries have deteriorated over the past 40 years, as a higher proportion of children are born in unsuccessful families, including ethnic minority and immigrant groups. Experimental evidence of a positive impact of early intervention on the children from unsuccessful families confirms many non-experimental evidences, which shows that in the absence of supportive family environment, child development and its results suffer. If society intervenes early enough, it can improve cognitive and unsuccessful families' social-emotional skills and health. Early intervention improves learning, reduces crime, fosters labor productivity and reduces the prevalence of teenage pregnancy. These interventions are seen as increasing the cost-benefit and rate of return. Aid programs to the disadvantaged families are under review, an early childhood intervention bring much greater economic returns than late intervention (student and teacher ratio reduction, rehabilitation of convicts, general adult literacy, increasing police funding). Investment into early childhood amounts much higher than in active labor market measures in Europe. Skill formation in life is dynamic in nature. Skill fosters new skills and motivation creates motivation. Motivation fosters skills and skills foster motivation. If the child is not motivated to learn and be interested at an early age are more likely his adult years will suffer failures in social and economic life. When society delays to interfere in an unsuccessful child's life, the treatment of failures becomes more expensive. Using knowledge about the importance of child's early years in shaping inequalities and developing workforce skills is necessary to reorient social policy.[21]

Definition for vulnerable or children at risk is not approved EU-wide, therefore used various methods to identify the most vulnerable groups of children. One such example is on February 2010 by TNS Qual + European Commission's Directorate General for Justice in all 27 EU Member States carried out a qualitative Eurobarometer survey of young people. The target groups of 170 were researched, object – rights of the child. Respondents of 15–17 year old were surveyed of different social, economic and ethnic groups, among them were the roma, Shiites and nomad children and children with special needs. One of the parts of the study was designed to identify the most vulnerable groups

---

Young Children // Big ideas for Children: InvestIng in Our Nation's Future. First Focus. 2009. P. 49–50.

21 Service infrastructure improvement possibilities seeking family welfare. Internet access: http://www.esparama.lt/es_parama_pletra/failai/ESFproduktai/2_SADM_tyrimas_paslaugos-seimos-gerovei_2013-07.pdf

of children. According to the survey results, the following main groups were identified:

(a) physical violence and children exposed to sexual abuse;
(b) children exposed to harassment – abuse;
(c) children living in poverty;
(d) children living in specialized institutions;
(e) the homeless, living on the street and formally not validated children;
(f) children with special needs;
(g) ethnic groups and immigrant children;
(h) children of roma, Shiites and nomad communities.[22]

The European Commission identifies several categories of vulnerable children. The first category – poverty and socially exclusion experiencing children – is defined on the basis of social, political and economic factors. European Commission[23] also notes that these groups of children often "tend to have drug abuse problems, lower learning achievements and poorer physical and mental health." The second category is impaired children. The third category of vulnerable children is children – victims of sexual exploitation or human trafficking. The fourth category is attributed to the asylum-seeking children and children who arrived unaccompanied in the EU. The fifth group of vulnerable children includes children – school dropouts, those who have only primary or lower education. Sixth vulnerable group of children is classified as Roma children. Seventh vulnerable group of children European Commission classifies are those escaped from home or institutions, and missing children. Eighth vulnerable group of children exclusion criteria is cyberbullying and school and seduction. At present, Lithuania material well-being – GDP – reached three-quarters of the EU average. Seventy-five percent of it consists of income: salaries, various allowances, income from independent economic activity, etc.[24]

---

22 Same.

23 European Commission: Communication from the Commission to the European Parliament, the Council, the European Economic and Social Committee and the Committee of the Regions: Agenda for the Rights of the Child, COM (2011) 60 final, 2011.

24 Income inequality research: Lithuania as country of paradoxes. Internet access: http://vz.lt/vadyba/2017/04/21/pajamu-nelygybes-tyrimas-lietuva--paradoksu-krastas#ixzz4ezoegBqY

Conclusions

We offer this approach for discussion by analyzing successful educational applications not only in the face of growing technological opportunities, but also in a closer look at community based initiatives by independent teachers, entrepreneurs, youth and NGOs that change the face and tradition of teaching. Analyze COVID 2019 we can argue that generation Z will be known in the future as the 'resilient' generation who learned at a sensitive age that it is possible to overcome crises without the entire life collapsing. With COVID 2019 lessons learned, the generation could grow up to become leaders who can really drive change and solve problems, when it is needed most. Generation Z values open to all world challenges and with mentors support we can find the best solutions for more safe and sustainable their life.

To create new standards and concepts, we need to enable the involvement of social partners, public spaces, active entrepreneurs, not only to facilitate sectoral interaction, but also to change attitudes and traditions. Smart education indicators need validated scenarios to change the traditional agendas of educational institutions, enabling them to engage and collaborate more actively with social partnerships for beginning social clustering. To encourage social innovation, more trust and better partnership between business, NGO and research sectors are needed.

Continuing with overall reform of the education, social, health and cultural systems balanced this reforms at all levels, addressing social partnership functions and skills mismatch. Social assistance, growing social justices and New Social Models have made our countries and social and education systems more sustainable. OECD suggests that further steps, for successful smart education, such as a rise in minimum pensions, positive discrimination tools for families with child poverty or special needs for children, are needed to reduce old-age and families' poverty and empowerment of social educational assistant and complex aid tools.

II Social Economy and Civic Engagement

2.1 Social Economy as a Platform of Smart Education

Smart Economy is a term that is old worldwide but new in Lithuania. This term has not been established yet by economists in our country, has not become common in the society. A concept of socioeconomics and activities related to it originated in Canadian province of Quebec. The first commitment of the federal government of Quebec in terms of socioeconomics became obvious in 1997. New economic tendencies occurred in Europe, Northern and Latin America and Africa, and today they exist all around the world. We perceive socioeconomics as pragmatic response to global economic and social challenges in times of crisis. The sector accounts for 2 million businesses, representing 10% of all European enterprises, and employ over 14 million paid employees (6.5% of the working population in the EU).[1] Social business, organizations of socioeconomics and individual initiators build business using community ideas and democratic operations and put individual and social priorities above pursuit of profit and autonomy of management. Socioeconomic organizations (social business) connect cooperatives, organizations of mutual aid, associations and funds as well as innovative business organizations. Manifestations of socioeconomics can be found in all sectors of the society and in all its levels. But it derives from the local (community) initiatives. These initiatives are especially active in the sectors of social security, social services, health, banking (credit unions), insurance, agriculture, consumer organizations, in the field of crafts, housing, initiatives of safe neighborhood, spare time, education, organizing of sports, production of healthy food. However, all these things can be described by phenomenon of smart education. Socioeconomic business forms a part of 10% of the total European business and there are up to two million enterprises which represent social business. Socioeconomics is referred to as economics of the third sector in the encyclopedia, that is, when business and authorities form the first and the second sectors. "Organizations created by civilians and operating in the structure managed in accordance with democratic principles and not pursuing for profit are the actors of socioeconomics. Their goal is to affect development of sustainable and harmonious society. Socioeconomic organization has to be convenient for people. Results of empirical studies clearly

1 Social Economy. In: https://s3platform.jrc.ec.europa.eu/social-economy

show cohesion between social competences/social capital and economic/social welfare. Social competences play a more and more important role in the developing knowledge-based society – knowledge personnel currently spend most of their working time searching for information, assessing it and communicating. Innovations are very important for knowledge-based economy, especially ability to intercept and improve them. In order to take advantage of abilities of their employees to develop innovations at work advanced enterprises implement modern forms of work organizing, that is, high-performance work systems. According to the data of international comparative studies, Lithuania lacks both social capital and social competences. There also exists a significant mismatch of staff qualifications, competence and demands of the labor market. As there exist coherences of social capital and social skills with economic growth, innovations and social welfare, their default has a negative impact on aforementioned factors. In order to achieve a high level of economic and social welfare it is necessary to raise a level of social capital and competences. This can be best achieved using the following two key measures of public policy: by promoting public activity (participation of people in the process of decisions making and in activity of non-governmental organizations and creation of environment which is suitable for than) and by improving a quality of education, by increasing its diversity and accessibility (by implementing and developing modern forms of training in the system of both formal and informal education). A person may change unsatisfactory situation. He is a master of himself, his family, school and the state. This shall apply in terms of education because all theories of social education are associated with a person-centered approach. The main principle of this methodology was developed in accordance with the ideas of gestalt psychologist Kurt Godlstein[2] extended by Carl Rogers[3] and Abraham Maslow[4] who have supplemented them with ideas of self-realization. A human is responsible when he is free and he is free when he is responsible (Richardson, 1985). The conception of a person-center is also relevant in

---

2 Goldstein, Kurt. The Organism: A Holistic Approach to Biology Derived from Pathological Data in Man. New York: Zone Books, 1995. ISBN 0942299973

3 Person-Centered Therapy (Rogerian Therapy). GoodTherapy.org. Last updated on the 7th of February 2015 [žiūrėta 2017 m. sausio 10 d.]. Access through the Internet: http://www.goodtherapy.org/learn-about-therapy/types/person-centered

4 Abraham Maslow. The pursuit of happiness. [date of review: the 10th of January 2017]. Access through the Internet: http://www.pursuit-of-happiness.org/history-of-happiness/abraham-maslow/

the social theories.[5] Analysis of scientific studies and welfare of life shows that social competences become more and more valuable in the modern economics. The available data suggest that social competences and social capital are clearly insufficient in Lithuania. Two main trends of actions, that is, promotion of public activity and improvement of educational system can be distinguished for the purpose of strengthening of social capital. It is recommended to promote a public activity by the following measures:

- by promoting volunteering and civic engagement;[6]
- by involving the members of the society in decision-making.
- by increasing awareness of the concept of social capital by involving it in strategic documents;
- by increasing transparency and accountability of policy;
- by developing socioeconomics;
- by reducing exclusion through provision of legal and civil assistance.

The following measures are recommended for improvement of educational system:

- to develop values during lifelong learning with the help of modern educational methods;
- to help develop and mobilize networks and mentoring which support cooperation and learning;
- to improve qualification of teachers;
- to involve societies in management of schools;
- to develop new forms of training as alternatives for a formal education, that is, such forms that improve social and civic competences.

In order to show that similar initiatives are already being implemented in other European countries, a small list of selected initiatives which strengthen a social capital in other countries is presented:

5 Marija Vyšniauskaitė, Prof. Dr. Giedrė Kvieskienė, Vytautas Kvieska, Tuomas Rauhansalo, How to educate a responsible student. Integrated education guidelines. Access through the Internet: http://www.llri.lt/wp-content/uploads/2017/02/kaip-ugdyti-socialiai-samoninga-moksleivi.pdf

6 Parissaki, M.; Humphreys, E. (2005) Regional social capital in Europe. European Foundation for the Improvement of Living and Working Conditions. Luxembourg: Office for Official Publications of the European Communities.

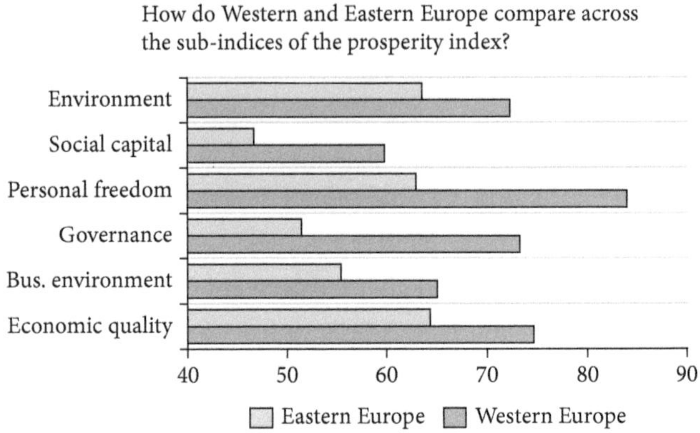

Fig. 18: Comparison of subindexes of the Western and Eastern European welfare

- In order to obtain a school-leaving certificate, Dutch students must have worked at least 30 hours for the sake of the community.
- The guidelines of the policy of national identity, civil society and integration, which were adopted in Latvia in 2011, induce development of both formal and informal civic education;
- The module "personal and social improvement" whereby, passingly, the country seeks to improve communication, teamwork, analytical and problem-solving skills is included in Maltese program of social sciences;
- The strategy of social capital development for a period of 2011–2020 is being currently developed in Poland;
- All Danish students over 16 years of age must spend 20 hours in organization of their choice that is outside the school (they must also engage in non-governmental activity and (or) volunteering);
- Volunteering is a part of the compulsory education program in Slovenia;
- Social Responsibility Development Center established in the Leeds School of Business of University of Colorado (USA) implements the undergraduate program in the field of social responsibility and other initiatives.[7]

7 Sekmokas, M., Social capital and social skills – Lithuania in the international context. Future business. 2013. Internet access: http://www.svak.lt/images/ateities_verslas-spauda_ok_web.pdf

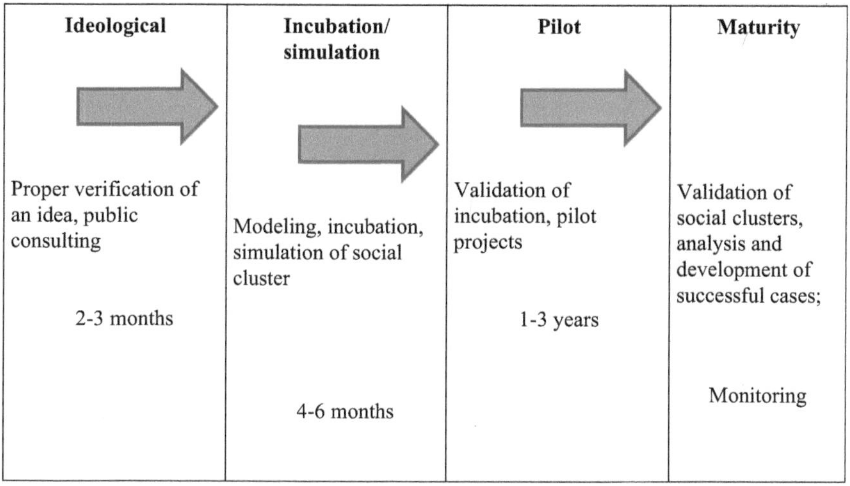

Fig. 19: Modeling of Stages of Social Cluster (Adapted by the author based on the recommendations of the World Forum on Social Entrepreneurship in South Korea)[a].

In order to analyze the existing experience of social partnership, social economy and social clustering required in the public sector the main focus was put on relations of NGO and state institutions in the field of provision of public services with emphasis on child welfare. The recent history of partnership between NGO and public sector was analyzed based on research and scientific studies performed in universities realizing that the existing situation is not spontaneous, and it was not created here and now but has arisen over a longer period of time and created traditions and expectations of NGO sector. The documents supporting partnership between NGO and the state were additionally collected from the Centre of Registers and Statistics Lithuania when cooperation between and reciprocal support of the state and NGO are high, though not necessarily equivalent by its influence and resources, during implementation of public policy and decision-making.

Integration of social education with its longest tradition in the United States of America and Canada, where the vast majority of training material and goals of the program are focused on integrated, interdisciplinary attitude, is closely related to enabling regional policy. The Adelaide Declaration (DETYA) urges to teach children at the schools to make their own decisions and assume responsibility on the issues related to moral, ethics and social justice, to build capacities of finding out the meaning of life and to be active

Picture 2: European Civic Society Representatives discussed poverty issues in The European Anti-Poverty Network 2016

and informed citizens committed to protect democratic principles and ideals. Japanese national education policy requires all students to study such integrated subjects as *people and industrial society*. It is stated in the Constitutions of the People's Republic of China that it is necessary to ensure civic education which would develop the national spirit, autonomy, moral, healthy body, management of scientific knowledge, enterprise and management of own life (Article 158). There is no separate social education course in the education programs of the Republic of South Africa. The Ministry of Education and Science of the RSA require to develop "respect for values of human rights, civic responsibility and respect for the environment" at the schools. "Social development" in other countries, such as Ghana, Zimbabwe, New Zealand, Czech Republic, United Kingdom, Hong Kong, other Asian and European countries, is the most commonly used in the context of faculties of social sciences, high schools and university historians, anthropologists, sociologists, economists and other social sciences.

Application of the principles of universal quality management in the regional policy[8]:

1. To ensure consistency of the goal. If the goal is unclear or constantly changing – development projects shall be suspended, new programs shall be developed, regions does not move anywhere. Thanks to the structured goal which is adequate to vision, values and mission and implemented in accordance with the PDCA (plan-do-check-act) cycle it is possible to move forward.
2. To adapt a new philosophy. A new philosophy is a continuous striving for perfection, whereas an old philosophy is seen as the method of attempts and mistakes. It is too expensive to use bad raw materials and to deal with consequences hoping (accidentally) to create a good product.
3. To reduce dependency from general inspection (control). Statistical quality evidences should be relied on instead (prevention of defects, not their search, should be better executed). Results must be forecasted considering input.
4. Not only a price should be considered. Accounting approach (optimization of input instead of activity) has unbalanced provision of many services provided in the local level (marginal examples – public transport in rural locations, rural culture centers, post office) – for this reason, the systems operate in unstable, unpredictable and low quality manner.
5. To look for problems and continuously improve all processes (kaizen). Improvement does not end with completion of the project or reform, it has to take place on a daily basis looking for any process which can be improved, must focus on systematic (not only local) changes of one person or group.
6. To conduct training. People can do wonderful things if they have knowledge.
7. To promote leadership (leader is an individual who realizes processes and helps others become better). Simple statistical analysis methods allowing early detection of errors and avoidance of negative outcomes correspond to one of the essential tools every leader should have.
8. To eliminate fear. Fear is a cancer which destroys any system from inside. Any other of 14 items will remain unimplemented if people will be afraid to submit suggestions, will hide information, will not cooperate and will delay changes.

8 Out of the Crisis. By W. Edwards Deming. Internet access: https://mitpress.mit.edu/books/out-crisis

9. To eliminate interdisciplinary barriers. The regional policy is based on horizontal cooperation and the Ministry of the Interior is probably the only ministry applying this principle on a continuous basis. The models of regional dimension are developed in the Ministry of the Interior – regional planning of projects, local development initiated by the community, integrated territorial investment – occurrence of these models in general and growth of funds is not a volitional decision of one institution but the result of negotiation process between managers of assignations, regions, municipalities and other partners not. However, reluctance to cooperate from the side of branch ministries remains a serious problem because thanks to elimination of barriers and openness it is possible to see (and admit) mistakes and prevent unilateral decision-making.
10. To eliminate meaningless slogans. Propagandist slogan which requires to work (live) better is seen as contrary to purpose based on objective factors and people who implement it have knowledge and abilities to effect it. Unfortunately, in case of regional policy, goals posed and measures available still are poorly linked (mostly because of reluctance to cooperate from the side of the Ministries of Economy, Agriculture, Education and Science which manage indispensable measures).
11. To remove target quotas (resources and output). The system of quotas (by fields, ministries and regions) applied for investment of the EU funds, state budget and other funds of the previous decade led to assimilation of these quotas. In this case, quotas of "objects attractive for tourism" or "kilometers of pipelines" and quotas of funds established for a region are little different from production plans applied during the Soviet period (when no attention was paid to quality and a real demand).
12. To eliminate barriers preventing people from enjoyment of their work and achievements. Performance of senseless tasks does not motivate. All people should realize their contribution and significance (and should be appreciated) – starting from the lowest links – a bus driver, social worker, librarian. . .
13. Improvement should be converted into a permanent and inclusive process. The region will grow together with growing experience, knowledge and expertise. Thanks to knowledge improved on a continuous basis, people will be able to improve their work, to create innovations, discover niches leading to success.
14. Everyone should be involved in the process of improvement. Activity result is always dependent on the weakest links.

A new model of activity of regional development councils is associated with the model of planning of the EU funds investment designed for elimination of competition between the projects of public sector. Regional development councils, which are responsible for selection of projects which are the most important for the region and assurance of compliance with the long-term strategic goals of the region by ensuring a long-term effect on competitiveness and quality of life in the region, are promoted too. Regional development council, municipalities and communities are involved in deciding on use of the EU investment in regions according to specific needs of a particular location. Greater support is allocated using models of community initiated local development that are proven to be right and are based on good European experience, by implementing the principle of leader[9] and thus deconcentrating management, strengthening administrative capacity of municipalities and communities, by ensuring a greater involvement of citizens into the process of use of the EU investment and in accordance with certain other principles:

1. Regional, local criterium (based on initiatives, innovations, social industries, creativity, partnership of the locals);
2. The priority of local initiatives (decisions are made based on decisions of the locals; mandatory conditions: partnership between private sector, public sector, science and NGO: legalized by contracts);
3. Partnership method and local activity group (LAG) (LAG – organizational unit that lives in the same location and that has legalized partnership intercourses);
4. Innovation (social innovations and SWOT of districts, strategies, decisions are made based on analysis of good examples)
5. Integrated method, examples of systematic (complex) problem solving;
6. Network activity and cooperation between locations (social partnership, social clustering: analysis of good experiences);
7. Funding and management at the local level.

The state (at least in the near future) will not be able to allocate support for regional policy to the same extent as support allocated from the EU. But ways of compensation of this misbalance using regulation, tax or other measures adapted for a particular case are sought. There is a plan to apply measures involving increase of management control, reduction of bureaucratism, greater empowerment of local government, communities and exploitation of their potential, and in the

9 LEADER program. Internet access: http://www.zmmc.lt/lt/leader-metodas.html

longer period of time – reform of the country's region management. A shift from sector (according to competence of ministry) to territorial way of investment (according to trends of regional specialization) is made by expanding competencies (capacities) of regional development councils and by ensuring accountability. The EU funds which are split between sectoral ministries (place-based approach) can be used more effectively using a territorial way of investment; accessibility of multi-faceted sources of project funding by subjects which develop local projects subjects is improved, resources of Lithuanian and foreign private participants of markets, finance institutions, initiatives of the European Commission for development of regional/local development are attracted; effective and sustainable (of partial/full return) model of regional development funding is developed by reducing dependance of municipalities on subsidies and grants (by solving at the same time problems of municipal borrowing).

Attraction of direct foreign investment to less developed regions from the third-rate activity (that was considered prospect-less) has become one of the main priorities in Public Institution Invest Lithuania. Thus, methodological assistance of the Ministry of the Interior provided to the specialists of Public Institution Invest Lithuania in interpreting of peculiarities of regions and regional policy, financial support (e.g., funding of young professionals program "Create Lithuania"), involvement of Public Institution Invest Lithuania in preparation of programs of development of integrated territories and cooperation with municipalities yielded results. A growing awareness can be noticed recently and specific suggestions are provided by other institutions – the Ministry of Foreign Affairs (economic diplomacy of regions), the Ministry of Environment (improvement of territory planning); the Ministry of National Defense (cooperation in consolidation of resources for local services provision) – that is, not only a coordinator is concerned about regional policy.

While preparing the Green book of regional policy, discussions with representatives of business and non-governmental organizations, academic community, regional development councils and ministries regarding the principles of a new regional policy (until February 2017) are held.

* The purpose: universally transparent and politically supported agreement regarding the principal provisions of regional policy. The aim is that all policy-makers and partners would have a general vision of regional policy (instead of separate visions of culture, economy, education, agriculture, the EU funds or other "regional policies").
* Consideration of the Law on Regional Development shall be renewed in the Seimas (during Spring session of 2017).

* Institute of Public councils shall be consolidated (the model of consultation and accountability of regional development councils)
* Representatives of social and economic partners (business, science, NGO) are granted up to 1/3 of votes in regional development councils.
* Preparation for implementation of regional policy in the modified NUTS system of regions.
* Modification of the methodology of regional development plans preparation is being prepared (until April 2017) (an essence – a set of tools shall be developed for implementations of other changes):
* The procedure of strategy planning shall be improved (regional specialization); establishment of pilot projects and projects of regional importance; the cycle of continuous planning (plan-do-check-act) shall be established instead of instantaneous planning and formal monitoring of quota; the procedure of consulting with the society shall be improved and etc.
* Simplification of structure of plans, automation of data processing, integrates statistical analysis and other (user-friendly) decisions in order the methodology would become simpler for the end user.
* Adaptation of the methodology of regional development plans preparation for planning of measures of all types (currently the system is optimized for the EU funds; other measures can be planned and managed, but this cannot be done conveniently);

An essence – mobilization of competencies in regions through joint work (and feedback to policy makers). Teams of experts with horizontal competences are formed of: the Ministry of the Interior: regional policy, management, local self-government; the Ministry of Environment: territory planning, urban development; the Ministry of Foreign Affairs: economic diplomacy, attracting of investment; the Ministry of Economy: small and medium-sized business; the Ministry of Agriculture (rural development), the Ministry of Finance (partnership between public and private sectors, financial engineering, public finances), Public Institution Invest Lithuania, science institutions (economics, management, engineering), social and economic partners together with local partners performing analysis of advantages of each region (relative) and preparing suggestions regarding trends of regional specialization. Campaign of involvement of the society must be carried simultaneously (publicity, sociological surveys, science research, measurement of satisfaction and awareness, workshops, etc.).

An essence in functional areas of regional planning – the same functional areas are used (defined through interrelations of residents) as for the systems of

Picture 3: For Smart Education we need Smart Environment (National Education Exhibition 2016, m Vilnius)

service sharing. However, a correct identification of financing gaps and needs (of regional specificity) is important in this case. Diverse intensity of investment measures (from full reversibility to grants) and a differentiated tax motivation system (direct foreign investment in free economic zone of Vilnius or Kaunas has little in common with municipal activity) should be applied realizing that different objective opportunities of private investment attraction exist in different parts of Lithuania.

We hope that modules developed will induce development of Local Development Fund (LDF): by combining public and by attracting private funding resources (resources of the Ministries from the EU funds, resources of Lithuanian and foreign private participants of financial market, international financial institutions (EIB, ERBP, etc., resources of bilateral and multilateral support of the European Economic Area, the program of Rural development, the program of Fisheries fund, the existing instruments of financial engineering (JESSICA, INVEGA guarantees), funds that return from financial instruments

Fig. 20: Sustainable development goals

of the EU structural funds during a period from 2007 to 2013, the state budget), that is, a form of funding of regional and local projects shall be changed from grants to a fully or partially returning investment.

A systematic review of the legal acts governs the planning system and integration of the principles of regional specialization into the methodology of strategic planning of the state (and the legal acts which govern the budget structure), that is, managers of assignations are deprived of the opportunity to make unilateral decisions regarding investment in the particular region – investment resources of the state can be assigned only for investment characterized by a clear national effect (for the projects of national importance) or for the general funding of the part of grants of the local development fund (in accordance with a clearly defined investment strategy of the fund and trends of regional specialization).

Conclusion

The social economy as smart education platform intends to deal with several challenges related to 3 M/D principal based on trust and social partnership in order to:

- empower educational institutions to became central actors for develop social clustering processes;
- create common value chains of social enterprises belonging to different sectors in communities;
- improve the partnership and sharing culture between schools, enterprises and other territorial actors, to empower STEM and innovative competences;
- provide local community actors with the necessary skills to generate and develop collaborative projects based on territorial potential;
- share and exchange best practices and experiences in the field of public policy to empower schools and support youth innovations and social enterprises.

2.2 Social Communication[1] and Mediation[2] as the Strategies for Clustering

Social communication is a term that refers to the social activity that is a form of the social interaction. Social communication is the way of transmitting information, knowledge, questions, instructions, commands, as well as rumors, gossip,

1 Communication Skills - How To Improve Communication Skills - 7 Unique Tips. Internet access: https://www.youtube.com/watch?v=mPRUNGGORDo Think Fast, Talk Smart: Communication Techniques Internet access: https://www.youtube.com/watch?v=HAnw168huqA

2 The Stages of Mediation: https://www.youtube.com/watch?v=KTONZIFm1t4 How Mediation work. Internet access: https://www.youtube.com/watch?v=6drwBHTMsK8 Mediation –How dos it work? Internet access: https://www.youtube.com/watch?v=6ayR5L0jfAM Mediation in action: role-playing workplace dispute resolution. Internet access: https://www.youtube.com/watch?v=LO3OMVWN8lk Mediating Employee Conflict | Conflict Resolution. Internet access: https://www.youtube.com/watch?v=oqFoy8ehzuc

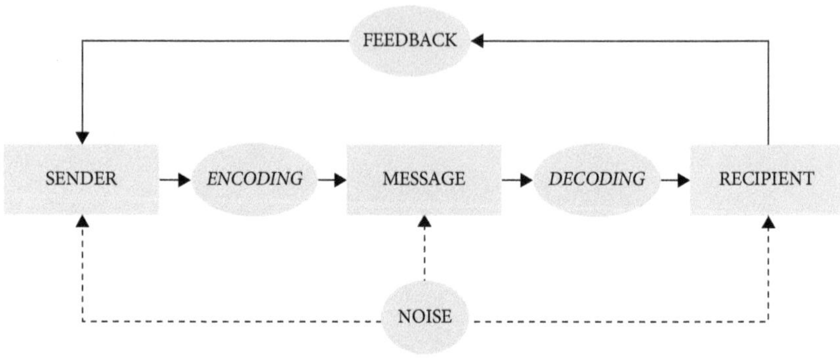

Fig. 21: Basic model of social communication. Basic Model of Social Communication https://managementmania.com/en/basic-model-of-social-communication

misinformation, hoax, whisper, etc. between the people. Social communication is a term that refers to the social activity that is a form of the social interaction. People in the organization, in its parts, teams and over the organization boundaries, they create different communication networks, which are a specific type of the social networks.

There are many models of the social communication, for example:

- Basic Model of Social Communication[3] is based on the communication model common in the field of information and communication technologies (ICT). This model has proved so useful that social psychologists "took it to mercy" and up to today, it is often used as a starting model. Model understands communication as a linear activity as follows:

  Sender (somehow) encodes the idea
  The idea takes the form of a message (spoken, written, nonverbal, graphic . . .);
  Recipient decodes the message and (somehow) understands it;
  Subsequently recipient provides feedback to the sender, whether he understood or misunderstood the message;
  Unfortunately, both the sender, the message (transmission) and the recipient are exposed to noise that blurs more or less the message;

3 Basic Model of Social Com https://managementmania.com/en/basic-model-of-social-communication

- Complex Models of Social Communication[4] are the Interactive approach understands communication as "ping-pong." Transactional approach sees communication as the exchange (transaction), as a complex game. The ball is now on one side, in next moment on the other side. Everyone is for a while a speaker, and also for a while a listener. They interact, however, the communication is quite orderly. Both parties are also speakers and listeners. Communication is often set so that both talk "over each other." New message on one side raises new message and response on the other side. And so over and over the communication content is getting richer and spreads (symbolically indicated by a spiral).

Social communication process is fraught with complexity and that the social communication process is also a social emotional process and communication process throw community, society, social networks. We respond emotionally (internally and perhaps externally) to those who are around us and how they treat us and others. Therefore, factors to consider are those related to social language and to social emotional processing. These include being able to read the context, responses related to nonverbal cues (i.e., situational cues), what one knows or doesn't know about the communication partner, the person's facial expression, body language, breathing patterns, and tone of voice, etc. At the same time, one must also consider one's own mood and the mood of the other person. The social communication process is synergistic and dynamic and constantly evolves with age.

Social communication skills[5] refer to all of the skills we need when using language to communicate and engage in conversations with other. Social communication encompasses the following skills:

- Using language for a range of functions, for example, to provide information; to question; to negotiate; to suggest and to clarify;
- Understanding conversational skills, for example,, starting and finishing conversations; maintaining a topic of conversation and taking turns in a conversation;

4 Complex model of Social Communication. Internet access: https://managementmania.com/en/complex-models-of-social-communication

5 Social Communication Skills. Internet access: http://www.hacw.nhs.uk/our-services/speech-language-therapy/childrens/social-communication-skills/

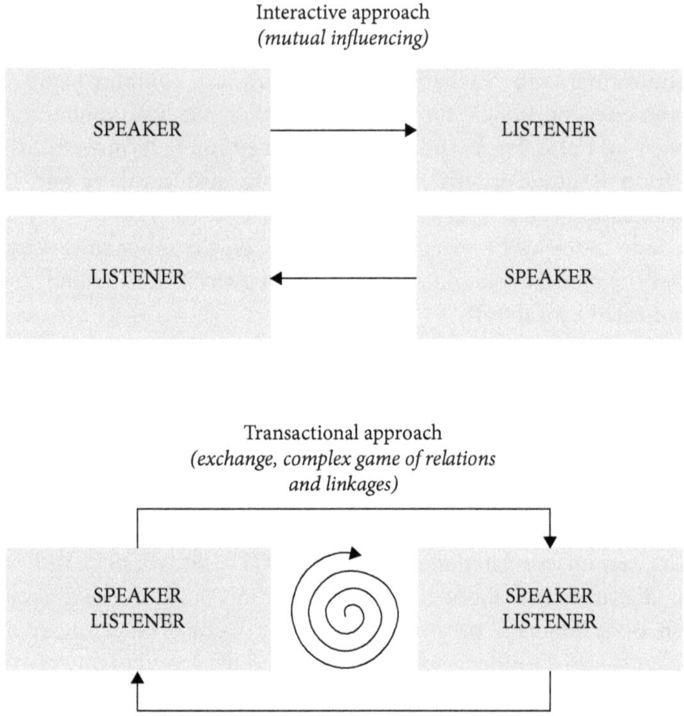

Fig. 22: Complex models of social communication. Complex model of Social Communication. Internet access: https://managementmania.com/en/complex-models-of-social-communication

- Understanding shared and assumed knowledge, that is, how much information the listener needs to understand;
- Understanding and using nonverbal communication skills, for example, eye contact, facial expression, gesture, proximity and distance;
- Understanding implied meaning.

What is Social Communication and Why is it Important? When we have a conversation there is a lot more to it than just saying words. We look each other in the eye, we take turns, we read facial expressions and body language, we pay attention to what our communication partner is paying attention to, and we stay on topic. We can read a lot from a person's tone of voice as well. When a child is having trouble learning about social communication, it is harder to

learn language, to join in conversations, and to play with other children. Your Language Express Speech-Language
Social communication difficulties can be associated with other developmental concerns. If your child has difficulties with social communication in most countries it is recommended for consultation Speech Therapist, Social Pedagogue, Psychologist or social education consultants, your family doctor. If there are multiple areas of concern (i.e. beyond just social communication) basic helps strategies (Social Pedagogy, Occupational Therapy and Speech Therapy) may well be recommended to address the functional areas of concern. How everyone can help:

- Smile at people when they smile at him
- Respond consistently when you say his name
- Pretend and play "make believe"
- Show an interest in other children
- Enjoy playing with you
- Make appropriate eye contact
- Show, give, share, and try to get your attention
- Ask for help when he needs it
- Greet familiar people

Helping Your Child Develop Social Communication Skills

Day-to-day interaction and play will develop children's social skills naturally. It takes time to learn the rules of interaction, but there are activities parents and caregivers can do which will help children understand the rules.

Eye Contact

- Get down to your child's eye level.
- Encourage your child to be in the same room and to face you when talking to you.
- Stop what you are doing; look at your child, listen, and respond. Then *wait* for your child's response. Don't rush communication. Children need lots of time to understand what they hear, and to think about what they want to say.

Turn-Taking

- Respond to your baby's noises and gestures and he will soon learn to make another noise or gesture for you.
- Play lots of turn taking games. Say "John's turn, mom's turn. . ."

- Practice sharing by taking turns with a toy. Games to help develop turn-taking skills
- Take turns putting the blocks on a tower until it collapses.
- Take turns blowing bubbles.
- Take turns choosing bedtime stories. Take turns turning the pages.
- Play board games with older children.
- Sing Nursery Rhymes. Leave out a word or two and wait for your child to fill it in.

Listening Skills

- Encourage your child to pause and listen, and then talk about the sounds you both hear.
- Turn off the television (iPad, phone, etc.) and talk and play with your child.
- Follow your child's lead and focus your attention on what interests her.

Shared Attention

- Follow your baby's gaze and focus on the same thing. Talk about what you both see. Eventually your baby will follow your gaze too.
- Go for a walk and point to, look at and talk about the things you see, hear, touch and smell.

Body Language and Gestures

- Use body language, facial expressions and gestures when you speak to your child. Your child will learn to "read" body language and to use it herself.

Kik Interactive[6], the creator of popular chat platform Kik, today announced it will unite with Rounds[7] the Israel-based communications platform, to boost Kik's expression and group video features, which launched in December 2016. The addition of Rounds expands Kik's international presence and will mark the opening of the company's first two product and engineering centers outside of Waterloo in Tel Aviv and Toronto. "Video is the natural evolution of chat and an integral part of expression and hanging out in real life. We saw Kik as the perfect partner to expand our reach to such a vast audience" said Dany Fishel, cofounder and CEO of Rounds. "It's rare when two companies share a similar culture and vision of connecting people through fun, meaningful experiences,

---

6 Kik. Internet access: https://www.kik.com/

7 Rounds is join kik. Internet access: http://rounds.com/www.rounds.com/blog/rounds-joins-kik/index.html

and we're eager to build the next great platform together." "The team at Rounds knows how to build products that make chat fun and entertaining for our teen audience," said Ted Livingston, founder and CEO of Kik. "The company has very strong product and engineering experience, and we're excited to join forces to achieve our goal of making chat the central hub for everyday life." Founded in 2009 by Dany Fishel and Ilan Leibovich, Rounds merges communication and entertainment together into a fun interactive video chat experience by enabling up to 12 friends to simultaneously video chat while experiencing activities together such as watching videos, playing games, sharing photos, and more. Rounds boasts more than 40 million users worldwide. All of Rounds employees will join Kik, bringing Kik's total headcount to 165. Kik's product and engineering center in Toronto will focus on chatbot platform development and will be based out of the MaRS Center. Both offices are planning to hire in the coming months.[8] Rounds offer instant group video chat on mobile. It's a live communication platform that makes it easier, faster and more fun to connect to friends and family. Using Rounds, friends can get together and send text messages, video chat with up to 12 participants and engage in activities while chatting – to watch videos, play games, share photos and much more. Rounds operates across networks, operating systems and devices. To date, Rounds has raised a total of $24 million funding from industry leading investors, including Sequoia Capital, Samsung Ventures, Rhodium, Verizon Ventures, DFJ's Tim Draper, and other private investors.[9]

Social mediation there are the field have debated the best method of training practitioners and whether mediation should have a professional education/training program. Moore describes three general mediator roles: social network mediators, authoritative mediators and independent mediators. Social network mediators are usually respected members of the community who have existing relationships with the parties. While not neutral, they are perceived as being fair. Social network mediators are generally concerned with maintaining stable long-term social relations. Generally, they remain involved with the parties after the negotiations, and will participate in implementing agreements. They are able to draw on social or peer pressure to enforce agreements. Authoritative mediators are individuals who are in some position of authority over the parties, such as a manager or director. There are a number of differences among authoritative mediators. They may be neutral as to the outcome, or may have vested

8 Connect the world throw chat: Internet access: https://www.kik.com/careers

9 Rounds is joing Kiek: Internet access: http://www.rounds.com

interests in achieving a particular settlement. Such mediators are generally able to use their authority to enforce agreements. Independent mediators are best defined by their neutrality and impartiality. Generally, they have no prior relationship to the parties, and they are hired by the joint decision of the parties. Independent mediators seek to help the parties develop voluntary, mutually acceptable solutions. The independent mediator model is most commonly used in Western countries; however, it is increasingly being used by other cultures as well.

Mediators of all types draw upon two broad classes of tactics: general tactics and contingent tactics. General tactics are kinds of interventions that mediators use in almost all conflicts. They include tactics for entering the dispute, analyzing the conflict, planning the mediation, identifying parties' interests, facilitating parties' negotiations and helping them generate proposals, drafting agreements and developing implementation plans. A key mediator activity is to identify the causes of the conflict, and to build a hypothesis as to how the conflict might be resolved. Contingent tactics are those used to handle the special problems which can arise during negotiations. Contingent tactics may be used to address such problems as value clashes, power imbalances, destructive patterns of interaction, communication problems, strong emotions, misinformation and differing analyses.

There are a number of factors which influence choice of mediation strategies. They include the stage of the conflict and the parties' abilities to resolve their dispute, the balance of power between the parties, which negotiation procedures are being used, how complex the issues are, and what the parties expect from the mediator. In crafting a mediation strategy, the mediator must decide on the level, target and focus of intervention. The mediator decides whether to concentrate on the level of general problem solving, or on a specific issue. She decides who she should be directing her activities toward. She also decides whether to focus on psychological, procedural, or substantive aspects of the conflict.

Mediators may enter disputes at the request of the parties, or by being appointed by an authoritative third party. The mediator has four main tasks at this stage. First, she must build credibility with the disputants. The parties must have confidence in the mediator personally, the mediator's parent organization, and in the mediation process itself. Second, the mediator must create a relatively close, comfortable relationship with the parties. Third, the parties must be educated about the mediation process, since it is ultimately the parties themselves who must resolve their dispute. This education also helps build confidence in the mediation process. Finally, after these steps have been taken, the mediator must secure the parties' commitment to mediation. This commitment may range from an informal oral agreement to a formal written contract, depending upon the situation.

Mediators may initiate contract with the parties in person or by phone or letter. Where there is some barrier to contracting parties, it may be helpful to have a secondary party introduce the mediator. The timing of a mediator's entry depends upon what type of intervention they plan to make. Interventions to collect data on the conflict can be undertaken at almost any time. The proper timing of problem solving interventions is disputed. Late intervention may allow the issues to develop more fully, and the disputants to have exhausted their desire and ability to continue the conflict. Early intervention may produce more cooperative, less coercive settlements and better relationships.

After the mediator has entered the dispute, the mediator and the parties must work together to decide what mediation strategy is best suited to their situation. Moore identifies six ways in which the mediator may assist the parties in choosing a mediation strategy.

- First, the mediator can help the parties to identify the interests at stake in the conflict, and to clarify their respective goals.
- Second, they can help the parties explore the range of possible, probable and acceptable outcomes. Outcomes may benefit both parties, cost both parties, or benefit one at the other's expense.
- Third, the mediator should describe the basic types of strategies for resolving disputes. The basic strategies are competition, avoidance, accommodation, negotiated compromise, and interest-based negotiation.
- Fourth, the mediator can help the parties to clarify the criteria that will guide their choice of strategies. Parties' choice of strategies will be shaped by time constraints, the nature of their current or desired relationship with the other party, their power and the parties' internal dynamics.
- Fifth, the mediator can assist the parties in weighing their options and reaching a decision.
- Finally, mediators can help the parties coordinate their strategies into a coherent, consistent approach to the dispute. Usually mediators and parties work together to analyze the conflict.

Conflict analysis begins with data collection directed by the mediator. Moore identifies six mediator activities with contribute to effective data collection:

- First, the mediator supplies a basic framework for understanding conflicts.

- Second, the mediator decides which data collection methods to use. Data collection methods available include direct observations, consultations with secondary sources such as maps or financial records, and interviews with the involved parties.
- Third, the mediator may delegate the actual data collection work. Interviews, in particular, will be more informative if the party identifies with the interviewer.
- Fourth, these data collection activities should be coordinated within a larger data collection strategy. This strategy should include an approach to identifying all the relevant parties to the conflict, a plan for the order and timing of the interviews, and a plan for building rapport with the parties.
- Fifth, the mediator should choose an interview approach. Interviews may be focused or unfocused, structured or non- structured, individual or joint.
- And finally the mediator must decide what sort of questions to ask in the interviews, and must exercise good listening skills during the interviews.

After the conflict data has been collected, it must be used to develop an interpretation of the dispute. The mediator's central task during this stage is to integrate and understand the elements of the dispute: people, dynamics, issues, and interests.[10] Timelines and case studies are two ways of integrating the data. Discrepant data must be discreetly verified. Mediators can begin to develop an interpretation of the conflict be separating unrealistic causes of the conflict from realistic causes. Unrealistic causes include stereotypes, miscommunication, confusion over the facts, and inappropriate competitive behavior. Realistic causes include competing interests, differing values, structural constraints, and disagreement over the collection or importance of data. In complex disputes, the mediator may choose to share her analysis with the parties.

Having secured the parties commitment and performed the conflict analysis, the mediator's next task is to develop a detailed mediation plan. "A mediation plan is a sequence of procedural steps initiated by the intervenor that will assist negotiators in exploring and reaching an agreement." In most cases, the mediator and the parties cooperate in developing a plan. A mediation plan should specify who will participate in the negotiations, whether outside- parties may be present, where negotiations will be held, and the physical arrangements of the negotiation space. The plan will also anticipate whether positional or interest-based negotiation procedures will be used, what interests and issues are most

---

10 Christopher Moore. The Mediation Process: Practical Strategies for Resolving Conflict. Internet access: https://www.colorado.edu/conflict/peace/moor7538.htm

important to the parties, what the parties' psychological conditions are, what deadlocks might arise and how they could be dealt with. The plan will establish ground rules for appropriate behavior in negotiations and should include a tentative agenda for the first meeting. Finally, the mediation plan includes a plan for educating the parties about mediation and negotiation processes.

Throughout the negotiation process, mediators work to build trust and foster cooperation. This activity is called conciliation. Conciliation is essentially an applied psychological tactic aimed at correcting perceptions, reducing unreasonable fears, and improving communication to an extent that permits reasonable discussion to take place and, in fact, makes rational bargaining possible. Moore describes five basic problems and the conciliation tactics which address those problems. Strong emotions can block agreements and inhibit the development of positive relationships. Usually mediators will allow the parties to vent their emotions in a controlled, safe setting. Venting may be inappropriate in conflicts with a history of violence or emotional escalation. In such cases, mediators structure the negotiations to suppress emotions, and may even resort to a kind of shuttle diplomacy. Stereotypes and misperceptions can be addressed by a four-step process. The mediator identifies the parties' perceptions, assesses their accuracy, assesses whether they help or hinder negotiations, and then helps the parties revise their inaccurate or negative perceptions.

Parties may question the legitimacy of other negotiators, of issues or of emotions. The legitimacy of negotiators may be established my revising misperceptions, and by explaining or revising how the negotiator was chosen to participate. Issues may be recast in more general or specific terms, may be redefined, or may be appealed to an independent authority. Challenges to the legitimacy of the other side's feelings can be addressed by exploring the reasons for those feelings, and pointing out that understanding another's feelings does not entail agreeing with them. When the parties' trust in each other is low, the mediator must suggest ways for the parties to build confidence in each other. Mediators encourage the parties to be clear and consistent, to follow through promptly on their promises, to expose small vulnerabilities to the other side, to accept penalties for breaches of faith and to refrain from making threats. Parties may also be given joint tasks to perform. Communication problems are common. Problems may occur with what, when, where, how, by whom or to whom a message is communicated. The mediator must determine where the problem is occurring, and then act to change the content, timing, setting, manner or agent of communication as appropriate. Mediators must also be attentive to nonverbal forms of communication and miscommunication.

The Mediation Process[11],[12] generally follow a common plan for the first mediation session. The session begins with an opening statement by the mediator. The mediator's opening statement introduces the mediator and the parties, defines the mediator's role, describes the mediation procedures including mediator neutrality, confidentiality, and the use of private caucuses, describes the meeting format, and suggests behavioral guidelines for the sessions. The mediator may then answer questions from the parties. Next, the parties each make an opening statement. Parties make a variety of different kinds of opening statements. They typically focus either on substantive interests and issues, on a combination of history, needs and positions, on the need for change or on some non-negotiable position. Occasionally they will focus on procedural concerns, or more rarely on psychological conditions.

Moore notes that "the most critical task for disputants at this stage is to maximize accurate information exchange. Mediators facilitate this exchange by using a number of communication tools. These tools include active listening, paraphrasing and restatement, summarization, and probing or clarifying questions. The mediator helps the parties to expand upon their messages, to structure their thoughts, to group and order similar ideas, and to separate complex issues' component issues. A positive emotional climate contributes to efficient communication. Mediators act to maintain a positive emotional climate by controlling or diffusing negative emotions or attacks, enforcing behavioral ground rules, and keeping parties focused on the issues.

Cultural differences may play a significant role in how negotiations begin. In cultures that rely primarily on social network mediators, the opening stage may involve much more informal conversation, and perhaps even food and drink. Authoritative mediators tend toward more formal arrangements. This formality is a necessary way of showing respect in more hierarchical cultures. High context, relatively homogenous cultures will spend less time explicitly specifying the process that negotiations will follow. Low context or heterogeneous parties will need to be more explicit. Communication styles also vary across cultures. Moore observes that what for one culture appears to be rudeness, interruption, and poor listening will be acceptable overlapping discussion in another.

---

11 Conflict Management (Funny animated), Internet access: https://www.youtube.com/watch?v=GnB3MpUfv_k TeamWork. Internet access: https://www.youtube.com/watch?v=1iayUeiP2gM

12 Christopher Moore. The Mediation Process: Practical Strategies for Resolving Conflict. Internet access: https://www.colorado.edu/conflict/peace/moor7538.htm

The next task is to set the negotiation agenda. The parties will need to identify the broad topic areas of their concerns, identify specific issues of contention, and decide on the order in which those issues will be discussed. Conflicts differ in the complexity and ambiguity of their issues. Generally conflicts are either interest based or value based. Value-based conflicts are less amenable to compromise and integration and so mediators should avoid describing disputes in terms of value differences when possible. The activity of defining the issues is called framing or reframing. Mediators help the parties to frame the issues in ways which will enable effective problem solving. Interest-based disputes can be reframed by either broadening or focusing the issues. Values-based disputes can be reframed either by reinterpreting the issues as interests, avoiding the value elements, or by appealing to broader shared values. Mediators tend to avoid using adversarial language, and instead use neutral terminology in describing issues and disputes.

Having identified the issues, the parties must now develop an agenda. Moore suggests a number of techniques for developing the agenda. Parties may address issues one at a time, in an ad hoc manner. The parties may alternate choosing issues to discuss. They may work from most to least important issues, or from easiest to more difficult issues. They may start by identifying which issues are key, and which issues are contingent on other agreements. They may package issues together to enable balanced trade-offs. Mediators are familiar with the advantages and drawbacks of these different approaches, and can advise the parties in agenda development.

Cultural differences can affect agenda development. Some aboriginal cultures prefer a storytelling approach to presenting the conflict's history and issues. Many cultures avoid direct confrontation, and will describe conflicts in indirect, oblique terms, or even in third-person terms. Such indirect cultures may prefer to begin by addressing areas of agreement, and by reinforcing existing relationships between the parties. Direct-dealing cultures prefer explicit enumeration of the issues.

Parties are very rarely able to give a clear or complete statement if their interests. Hence one of the mediator's important tasks is to uncover and clarify the parties hidden interests. Parties may be unclear on their interests for a number of reasons. External factors such as advice from friends or traditional explanations may cause parties to misperceive their interests. Parties may intentionally hide their interests in an attempt to gain a negotiating advantage. Parties also tend to confuse their positions with their interests. Often parties are not used to thinking in terms of underlying interests. Mediators should help to cultivate a positive attitude toward interest discovery in the parties.

There are two types of procedures for discovering interests: indirect and direct. Indirect procedures include testing, and hypothetical modeling. Mediators test for interests by listening to a party's statements, tentatively identifying the interest expressed, and then expressing that interest back to the party for further feedback. In hypothetical modeling, the party ranks a number of hypothetical settlements in terms of satisfaction. Direct methods include questioning the parties, and brainstorming. Parties may try to bluff, or misrepresent their interests. Mediators must be alert for such bluffs, and try to call them. Mediators may use persuasion or rationalization to induce a party to abandon their bluff and present a more accurate picture of their interests. The mediator must also encourage each side in the dispute to acknowledge the other side's interests, even if they do not agree with those interests. Finally, the mediator should incorporate both parties interests into a joint problem statement. A joint problem statement "enables negotiators to commit to work on a common problem because they believe that their needs will be respected, if not met by, the solutions that will be developed.

The parties are now ready to begin generating settlement options. Often the parties must be convinced of the need to generate a range of different settlement options. For instance, a party may believe that they already have the best solution, and merely need to persuade the other side to accept it. Settlement options may address either specific issues or general principles. There are two basic procedures for generating options: positional bargaining and interest-based bargaining. Interest-based bargaining is more likely to produce integrated, or win- win, outcomes, and so it is the preferred approach. Here again however, cultural differences are significant.

Moore lists a number of specific procedures for generating settlement options. Where the parties have an established relationship, they may proceed by ratifying the status quo, that is, by specifying which elements of their current relationship they would like to continue in the future. Parties may set boundaries for acceptable options by developing objective standards. The parties may generate options by open discussion, either in small groups or all together, or may use the more structured brainstorming procedure. Small groups may be asked to develop hypothetical settlement scenarios. The parties may consider settlement agreements from other similar disputes. They may identify linked issues, and explore possible trade-offs. Or they may consider various comprehensive package settlements. They may proceed by drafting and redrafting a single settlement proposal. Rather than consider substantive settlements, the parties may seek an agreement on a procedure for making decisions. Bringing in outside experts or other resources may help to broaden the parties' perspectives on the dispute and on the range of possible solutions. Research has found that stronger agreements tend to be substantive, comprehensive, permanent, detailed, non-conditional and binding. Agreements that are procedural, partial, provisional, abstract, contingent or non-binding tend to be weaker.

Having generated a range of settlement options, the parties must next evaluate those options. Moore describes two evaluative tools: the settlement range and the best alternative to a negotiated agreement (BATNA). A party's settlement range is defined by the range from target point to resistance point. The target point is their optimal outcome. The resistance point is set by those outcomes which are too costly or not beneficial enough to accept, that is, below the party's BATNA. When the parties' settlement ranges overlap, there is a range of possible mutually acceptable settlements available to the parties. Interest-based bargainers will seek the solution which provides maximum mutual benefits. When the parties' settlement ranges have no overlap, there are no mutually acceptable settlement options. In such cases, the parties may be encouraged to shift their target or resistance points, or the issues available for trade-off may be expanded. Mediators can help the parties to identify and understand their settlement ranges, and can deflate the parties' inflated expectations. Assisting the parties in identifying their BATNAs can also help to close any gap between the parties settlement ranges.

After evaluating the options, the parties enter the final bargaining stage, where they negotiate to reach an agreement. Occasionally an agreement will have emerged directly out of the evaluation stage. Usually however there is still a range of potential agreements, some areas of disagreement, and a number of details to be worked out after the evaluation stage. Moore identifies four basic strategies mediators use during the final bargaining stage. In the incremental convergence strategy, the parties each make small concessions until they reach a mutually acceptable compromise. When positional bargainers use this strategy, the mediator's main task is to make the parties comfortable with making concessions. In the leap-to-agreement, the parties engage in some preliminary bargaining, but then leap directly to accepting a comprehensive proposal. In the agreement-in-principle strategy, the parties first seek agreement on general principles, and then seek to apply those principles to the situation at hand. This strategy is particularly helpful in preventing negotiation deadlocks. When the parties cannot reach a substantive agreement, they may use a procedural strategy. Process decisions there are, when parties make to resolve disputes without directly deciding the issue. Generally, the process determines a substantive answer.

Deadlines can play an important role in bringing the final bargaining to a close. Deadlines may be internally or externally imposed, actual or artificial, rigid or flexible, with or without consequences, explicit or vague. Mediators can help make the parties aware of existing deadlines, or help parties establish deadlines as needed. Mediators can also help the parties avoid harmful uses of deadlines, such as using deadlines as threats or allowing deadlines to generate false agreement motivation. Cultural attitudes toward time affect the use of deadlines. Some cultures view time as a valuable and limited resource, while others have a much more leisurely attitude toward time.

After final bargaining, the rough agreement must be formalized and an implementation and monitoring plan developed. To be effective the agreement must be put into action, and so a well-designed implementation plan is crucial. Immediate, self-executing agreements are easier to implement. However, many settlements will require parties to act over an extended period of time. Implementation plans will be more successful when they include criteria for measuring performance, methods for measuring, enforcing compliance, general and specific implementation steps and procedures for managing future changes or conflicts. Some cultures incorporate implementation plans into the settlement itself, while others approach implementation as a separate set of issues.

Agreement formalization provides an important symbolic end to the conflict and increases the parties' commitment to the agreement. Formalization may take the form of verbal promises or written agreements, and may be public or private. When the parties' voluntary compliance is not sufficient, structural factors and external authority may be used to enforce compliance. For example, the settlement may be made legally binding and subject to judicial oversight, or the parties may be required to post performance bonds.

Moore concludes this section by discussing five common types of contingent mediator tactics used for dealing with special situations. These include caucusing, mediator pressure, power-relation management, negotiation teams, and constituency management. Caucusing allows the mediator to communicate with one party privately and separately from the other parties. The mediators may use caucuses for a number of purposes, including to discuss confidential information, to interrupt negative emotional dynamics between the parties, to identify the party's settlement range, or to discuss the party's negotiation strategy. Mediators are able to influence negotiations by means of their control over the process, setting and timing of negotiations, by their involvement in the parties' communication and relationships, and by their access to both the parties information and their ability to bring in outside experts. Generally mediators exercise their power in order to move the parties toward a solution, while remaining neutral regarding the content of the solution.

Mediation is most effective when the parties have symmetrical power relations. When the parties have unequal power, the mediator may need to balance their powers. In balancing the parties' power "the mediator provides the necessary power underpinnings to the weaker negotiator: information, advise, friendship, she may reduce those of the stronger. Complex, multiparty disputes may involve the use of negotiation teams, rather individual negotiators. When negotiation teams are present, mediators will need to pay attention to the interpersonal dynamics within as well as between the parties. Mediators may seek to influence the choice of negotiation team members and the teams' decision-making strategies. In some cases, the negotiated agreement must finally be ratified by the negotiators' constituencies. In such cases, the mediator will need to assist the negotiators in communicating with their constituents and in organizing constituent support for any agreement. The mediator may intervene directly with the constituency groups to educate them regarding the negotiation process, and to reassure them that their negotiation representatives have worked hard to achieve the best solution possible.

Conclusions

Mediation is a broad and rapidly growing field. There are now many professional mediation associations, focusing on areas ranging from divorce mediation to public policy mediation. These associations generally seek to develop and improve the practice of mediation by developing ethical standards, by offering mediation training, and by promoting the use of mediation. Moore reviews a number of such mediation associations and their activities. Generally, mediation codes of ethics require the mediator to be neutral and impartial, to avoid conflicts of interest, to obtain informed consent for their involvement from the parties, to maintain confidentiality, to reject cases which are beyond their expertise, and to be truthful in advertising their services and fees. The various associations' codes vary in their specifics. Research into mediation theory and practice has led to a proliferation of high-quality mediation training programs. Moore describes some of the basic elements of mediator training programs, including applicant screening procedures, training formats and methodologies, instructor qualifications, and criteria to guide the would-be mediator in selecting a program.

As the mediation field has developed, there has been an increasing trend in the public and judicial spheres toward using mediated negotiation as an alternative to more traditional means of dispute resolution. Legislation has been passed at the federal and state levels encouraging the use of mediation and executive orders have also mandated the development of alternative dispute resolution procedures in U. S. government agencies where appropriate.

Finally, Moore sketches five developments which must occur if mediation is to become even more widely utilized, and seven areas where the practice of mediation itself may be further developed. In order for mediation to be more widely used, the public must be better educated about it, more research into mediation needs to be done, mediation must be more fully institutionalized, mediation services must be more fully funded and new applications for mediation must be discovered. Moore suggests seven such new areas of application: ethnic conflicts, intractable conflicts, violent conflicts, education for the development of nonviolent cultures, dispute systems design, public governance and global environmental issues

2.3 Smart Communities and Regions for Smart Education

Regional Policy[1],[2] – is the public policy, aimed at the redistribution of public investments in favor of less developed regions. The aim of the regional policy is the reduction of regional socioeconomic disparities within the country. Horizontal measures of national regional policy are implemented in all countries territory via regional development plans, national strategies and programs, local and regional initiatives.

We can distinguish the European Union (EU) and Lithuanian national regional policy. The EU's regional policy measures to reduce the level of prosperity (as measured by GDP per capita) the differences between the most and least developed regions in the EU. Lithuania is considered as a single region, the closer to the average EU level of economic development is achieved. The EU's regional policy instruments – the Structural Funds. Strategic provisions of Lithuanian national regional policy are incorporated into EU structural support programing documents: Strategy and Operational Programs in national strategies. This allows to use EU support for national regional policy. Smart cities are groups of people who are working together, making use of energy supplies, raw materials, social services, financing and changes in technology to facilitate a flexible, excellence life and a sustainable regional economic development. They are six main approaches of a smart community, they include smart (circle, sharing) economy, smart people, smart governance, smart mobility. These aspects become possible through the making of innovations, planning and strategies to meet up with the needs of the citizens. Innovation on the other hand is the remarkable enhancement on goods and services which is as a result of technological improvement. Still, in order to achieve innovations, manpower also known as the traditional method, investments in research, development (R&D)[3] and education are required to make the most of innovations. However, there are other factors that powered innovation; social capital which is gotten from our social environment; findings for R&D and the institutional

---

1 Samprata, tikslas, uždaviniai. Internet access: http://www.nrp.vrm.lt/lt/regionines-politikos-formavimas-ir-igyvendinimas/samprata-tikslas-uzdaviniai/6

2 Lithuanian National Regional Policy. Internet access: http://www.nrp.vrm.lt/en

3 Anneli Kaasa, Helje Kaldaru, Eve Parts. Social Capital and Institutional quality as factors of innovations: 2007

environment, these factors helps to reduce cost and the risks involved in innovation. Many communities and schools are exploring the smart socialization concept. They may be installing municipal wireless networks; implementing e-government initiatives by providing access to city departments and initiatives through websites; integrating public transportation with Smart Transportation Systems; or developing ways to cut their carbon footprints and reduce the amount of recyclables that are consigned to the trash heap. In monograph we try clarify the meaning of the term "smart socialization and smart education" through an examination of more than 50 actual smart communities' projects around the world. It also provides insights and data from a range of ABI Research studies that cover many of the technologies that can be used to make a municipality and its various agents more communicative and responsive to its citizenry, while allowing its transportation and other infrastructures to use energy more efficiently and cost-effectively. Europeans Communities are pleasantly becoming a central point of interest in their societies and in the world at large, principally because of the process of economic development and urbanization in Europe countries, and the increased in the utilization of resources. In other to meet up with the depends of the citizens, nevertheless, there is a need to bring a transformation in the cities and in the underdeveloped areas by developing them to become smarter, consequently this need has to be achieved in a smart way so that they can be called smart cities. Defining the Smart Community concept is not easy given that "smart" is a relative term. Smarter Communities underscore the idea that a communities' improvement and forward-looking development should be benchmarked against itself (via improvements to its technological systems) and avoids the implication that any community can became a leader of socioeducation grow in the region. Many communities around the world are exploring the Smart Community concept. These communities may be installing municipal wireless networks, creating social clusters, implementing e-government initiatives by providing access to city departments and initiatives through websites, adding smart and sharing a transportation features to public transportation systems, or developing ways to cut their carbon footprints and reduce the amount of recyclables consigned to the trash heap. While definitions vary, it is generally agreed that a Smart Community can be judged on the basis of six dimensional axes:

- Smart (sharing, circle) Economy
- Smart Environment
- Smart Governance
- Smart Lifestyle

- Smart Transportation
- Smart Education

Again, European Innovation Partnership on Smart Cities & Communities search for an extensively increase in the industrial sector for the smart city resolution; amalgamate technologies from Energy; Information and Communication Technologies (ICT); finally, transport system.[4] To make every effort to bring more benefit to Europe countries; a noteworthy development in the life of European citizens'; an improvement in the level of contest of European industrial section, innovative of Small and Medium Sized Entrepreneur and a strong involvement in the Europe 2020 vision for energy and climate change. All this will be accomplishing through integration, changes, sustainable Smart City resolution, in areas where energy production, distribution of resources, transport mobility, information and communication technologies are closely linked. The European Innovation Partnership aims to triumph over the challenges that get in the way in achieving smart European cities. Adding together, to create innovation Strategic by Implementing Plans that as well will create value for Europe cities, to generate more successful knowledge allocation and also to institute partnerships among industry, innovation of SMEs, in European cities and other local stakeholders crossways European borders.[5] Citizens are consequently involved in these developmental processes but they are absent to motivate and make powerful to contributions. Hence, if they participate the strategies put in place will become more successful; on the other hand, measures and tools are put in place to furnish citizens to be actors in smart city developmental systems to ensure that they are educated, forced to act correctly, hands-on and participative, in this ongoing innovation process. In the long run due to their participations, creativity and entrepreneurship could be mammoth, offering massive unexploited prospective. Furthermore, the ICTs is another hindrance to this process of innovation, the most significant is the Internet, through the use of smart phones, because of the enthusiasm open toward new citizen programs that may not fit in with the current organizational system.[6] There is a Strategic Implementation Plan (SIP), which is a product of the work of a High-Level collection, contributions from the smart cities stakeholder platform and their supporting Sherpa group. The plan concentrates on some unambiguous, straight up areas at the moment in order

4 European Innovation Partnership on Smart Cities and Communities, 2013
5 European Innovation Partnership on Smart Cities and Communities, 2013
6 Internet access: http//www.Eu-smartcities.eu/

to focus on energy and to put on some propulsion together: Sustainability of Urban Mobility, energies, public transportations, resourceful logistics, and planning, improving the energy competence of buildings and districts, mounting the allocation of renewable energy sources and the livability of the communities; Integrating ICT and transport to improve the effectiveness and sustainable cities all over Europe.[7] Moreover, builders of smart communities are always organized, competitive, benevolent to build an energetic society where citizens will accept to work together with the states to achieve smarter societies. The following points are used to make a smart community:

- **Independent, energetic society, healthy and creative:** the states need to build its own society and environment. The citizens as well needs to be creative and think critically, the states should be able to ensure good public health and citizen's participation because a healthy life leads to the decrease in diseases.
- **Consolidated, dignified, responsible for a common fate brought by the citizens:** the state need to build up cultural and political consciousness in the country to endorse national identity, stress on positive historical experience; to reinforce historical self-awareness and self-esteem to support different educational and artistic appearance, also, to manage national study programs as the basis for humanistic education, training of generic cultural competencies; moreover, to generate a vivacious public space for information by promoting social and responsible media for information in the public area; to introduce media literacy programs in educational organization; to respect the autonomy of the press, and encourage self-regulation processes in the media, to create more job in order to limit migration; to create constructive working conditions for workers.
- **Creating an effective system of learning:** to successfully adapt information communication technologies (ICT) to ensure the acquisition and expansion of information and skills necessary for an energetic society. To develop national programs to let loose learners' prospects and talents, as well as academic mobility; to support non-formal education for gifted children and to bring together the country's best teachers, science, culture and sports experts, to be a magnet for overseas expert; also to build an environment constructive for science and research and to create a world-class studies and research center, to mobilizing the best scientific and teaching potential

7 An Energy Policy for Europe (2007) Internet Access: https://www.ifpenergiesnouvelles.com/article/smart-city-energy-challenges-facing-sustainable-cities

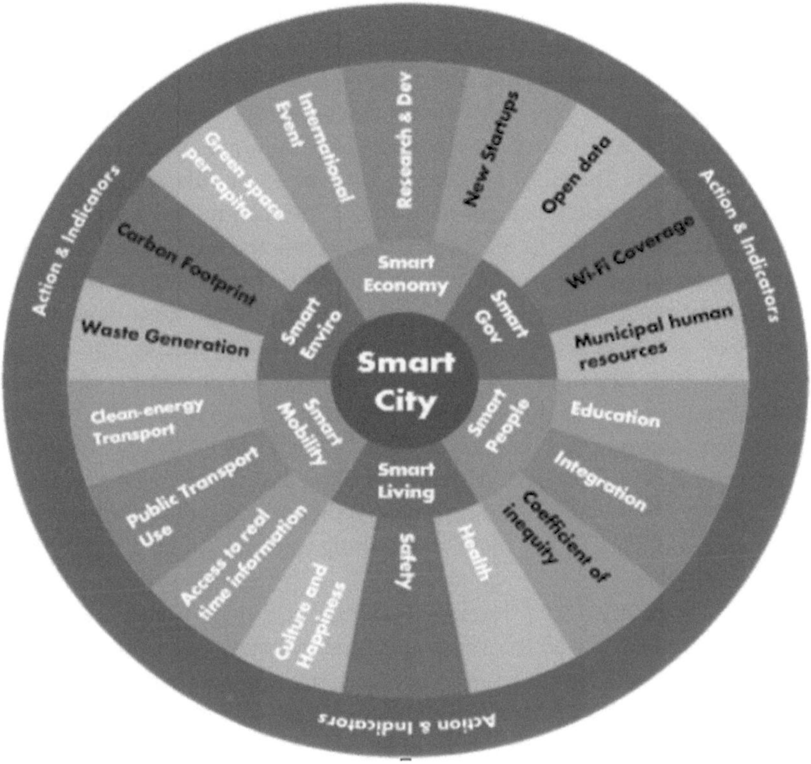

Fig. 23: Smart city

which will bring together study opening in the interdisciplinary network for research and development, and opening up the research infrastructure for business-science dealings; moreover, to expand a wealthy artistic milieu by investing in the expansion and amalgamation of public cultural institutions, and supporting public contribution in artistic growth; finally, to ease cultural vitality, principally during global cultural connections and worldwide artist mobility.

Furthermore, innovation has a great duty as far as economic growth and development are concern. However, it is essential to identify different aspects, which decide on the intensity of innovation in different regions and countries of the world, more emphasis will be on European countries and the old member states of the EU, countries with no communist setting and transition countries.

The figure represents the Cohen's Smart Cities Wheel, a model that offers six intelligence dimensions to consider in all smart cities: smart economy, smart government, smart people, smart living, smart mobility and smart environment. Text in white represents topics discussed on Twitter. Text in black represents topics not discussed on Twitter.[8] Open Data, are closely related with the concepts above. Therefore, we are not surprised that this argument is not treated. Finally it is worth mentioning that a city cannot be called "smart" if we do not take iniquity into account. Gini coefficient of inequity, which is defined by the Italian statistician and sociologist Corrado Gini, is measuring the inequality of income or wealth. We think that this technical data should become part of the education of citizens through social media. The market tends to produce cities that are not the reflection of consumer preferences incentive build vertically. If the municipal human resources gave attention to citizens' opinion expressed on social media, the buildings developer would be bound by the approval process. Although it is too early to tell what new forms of participation will exist, it seems desirable that smart and aware people on social media could be the key tool for spreading information favoring the participation of others.

Finland, for example, is a developed country and they receive the highest incomes. It is also one of the most educated country, because of the safe learning environment for students and of teachers. And another good example is Singapore. Even though it is one of the smallest countries in the world it has a dynamic economy, with low level of taxes and low level of corruption, Singapore is one of the foremost city that applies ICT strategies and smart policies in the urban areas. IT education, IT infrastructure and IT economy are the three pillars of the Singapore IT strategy (Karadag, Tunc, 2013). With plans put in place to make it to be a success, Singapore is the center for international trade technology and transportation. Also, because of the growing population and urbanization in Turkey, the state through the Ministry of Environment and City Planning has put in place majors' strategies plans and regulations to targeting the 2023 vision. However, the best European countries to live in are Finland, Switzerland, Sweden, Australia, Luxembourg, Norway, Canada, the Netherlands, Japan, and Denmark. United States.[9] A countries welfare,

8 Smart City Wheel Model. Internet access: http://chorally.com/learn-impact-smartcity-using-social-networks-analysis/

9 Giedre Kvieskiene, Guidelines on social cluster and social innovation competence model 2017

happiness, civic index confirms in part that investments in education, social and economic innovation and investment in science and development leads to changes in economic and quality of life (Giedre Kvieskiene 2017). In March 2011, The European Commission launched the Social Innovation Europe initiative; it was intending to help social innovators to develop goods and services, to find experts to develop social innovation and new businesses, latest working methods that will impact organizations, customers, and the society at large. In order to achieve the innovation goals, the states and non-governmental organizations have been carrying out scientific and technological research and development (R & D) to improve on innovation. Also, to achieve the Europe 2020 developmental visions measures have been put in place to mobilize more participants, networks and consultants. Innovation that expresses new social structure formation in societies is directly related also to public culture and mentality changes and the growth of social capital (Giedre Kvieskiene, 2017).

Most European economy are mostly affected by a decline in economic and finance. Therefore, there is a need for innovation to assist in solving this problem by promoting competitions, improvement in technology and businesses, which will therefore increase research institutions, economic growth, create an innovative society, encouraged foreign investment, promote business network and innovation network, to build a creative society and improved the living standards of the citizens. For example, there has been a lot of improvement in the Lithuanian economy because of the improvement in research. In addition, to overcome the financial crisis more attention has been given to social inclusion and the idea to promote social innovation, and so, this have help to close the gap between the rich and poor in the country. Hence, most residents in Lithuanian are faced with poverty and exclusion making the country to be at the fourth place from the bottom in the European Union (Giedre Kvieskiene, 2017). Although it is too early to tell what new forms of participation will exist, it seems desirable that smart and aware people on social media could be the key tool for spreading information favoring the participation of others. On the other hand, innovation on green space, mobility and smart living are well discussed on Twitter. One of the challenge regards the solution in climate change and environmental care. The school of energy of the University of Trento (Italy) organized the first edition of the IEEE Italy Section School on Future Energy Systems. The social Media gave a lot of attention on it, where the goal is to provide Ph.D. Students with state-of-the-art skills in this exiting and broad research field. Another key priority was related to smart care. Bolzano (Italy) has developed a project called "Abitare Sicuri" (live safely) addressed to the elderly. The aim is to promote the philosophy of dignity and independence as it is about equipment and services. The efficiency of such a system is not only based on technologies, but also integrating it with the involvement of all citizens. The great interest shown on the social media confirms that the experiment is on the right track.

In summary, living smart is the goal of most countries, though other countries might not have the resources to make changes, yet they work hard to make sure that citizens enjoy some reward of economic development. Nevertheless, some countries have all the resources to be smart so they put all the resources together to make better innovations for better living. Social partnership and Research are important tools for economic innovation and social innovation because they help to bring in new ideas, workplaces, and improvement of quality of life and more qualitative and more efficient social solutions.

Regional differences in the formation and dynamics determined by a variety of both internal and external factors, such as geographical location, natural resources, transportation network, access to major transportation corridors, urbanization, infrastructure development level, energy prices and availability, general domestic and foreign macroeconomic factors (economic recession, the European debt crisis, the situation on export markets and even the labor market situation in countries such as Great Britain, Norway, Ireland). Similarly, large economic operators and the general public's economic policies (employment, investment, innovation, basic infrastructure, energy regulation and development). In a small and open country, which is Lithuania, many external factors have a direct impact that cannot be removed and can only mitigate their negative effects. Therefore, regional policy Lithuania is compensatory in nature, aimed at reducing the negative impact of external factors and focused on the differences between regions and within them is minimized.

Given limited resources, Lithuania's regional policy is implemented in accordance with two basic principles:

- Subsidiarity: decisions should be taken at the level at which they are effective. When planning investments for regional development, partners must participate in regional institutions. In this way, local and regional partners are actively involved in investment planning and implementation process and help them steer a way that best meet local needs and opportunities.
- Concentration: mobilizing resources for smaller quantity of the target areas, the effect is greater.

National regional policy objectives of the tasks:

1) distinguishing the target areas to identify and implement their development goals and objectives;

2) addressing individual industries (sector) specific regional and local issues;
3) creating the conditions for long-term steady and sustainable development of all regions.

In order to effectively and successfully implement the regional policy in 2014–2020, EU funding developed a new method – an integrated territorial investment (ITI). The method is based on the fact that the city is viewed as a whole, which must be managed and developed as a comprehensive and specific "sectoral" issue, dealt with only as parts of the whole. The new method allows ITI to solve complex social (unemployment, migration, social exclusion), economic (inefficient, unsustainable agriculture), environmental (air, soil, water and visual pollution) and climate change, to coordinate public, private and community investment initiatives.

All municipalities Lithuania has adopted an integrated territorial development programs for the period 2014–2020 period . The integrated sites in ten regions and the five major cities of the development of the total will be directed to more than 700 million euro of EU investments.

Regional teams of experts shall be formed, the representatives of the ministries shall be replaced with those attributed by the implementing institutions – the center of competencies shall be formed (regional chamber): (CPVA (the division of public and private partnership and divisions performing functions of the implementing institution), other agencies of the EU funds, VIPA, Public Institution Invest Lithuania, Public Institution Enterprise Lithuania, the National Paying Agency, National Land Service, State Territorial Planning and Construction Inspectorate, the Chamber of Agriculture, Chamber of Commerce, Industry and Crafts or subdivisions of other umbrella business associations, etc. – composition depends on a specific region, regional development council shall coordinate the activity), that is, it becomes possible to take advantage of a professional consulting service at the local level – assistance and consulting on legal and financial matters as well as on issues of strategic and project management, public and private partnership (PPP), debt restructuring (in the future), attraction of foreign investment and other issues are provided in one place, documents required for implementation of projects can be prepared expeditiously (e.g., special plans).

The main (proximate) tasks of regional chamber:

- Idea testing – by preparing and implementing pilot projects (PPP, attraction of foreign investment, innovative projects of cities development, business-science cooperation);

- Preparation and coordination of projects, which go beyond administrative boundaries/competence of one institution – from preparation of conception and attraction of partners to the final achievement of result. A steady flow of high quality and high-impact projects is ensured in due course (not limited to use of a specific funding source or an area of interest of one institution).
- Assistance to project preparing and implementing municipalities, business and other entities striving to raise the project quality.
- Assistance during preparation (update) of the plans of regional development (together with mid-term review of the EU funds).
- The consulting team for any investor who is settling up in the region (regardless of whether the capital is of domestic or foreign nature).
- Analytical assistance for central authorities (the Ministry of Finance, the Ministry of the Interior) in development of the local development fund (LDF). The essential difference between the LDF and the previously applied measures of financial engineering: the projects funded should be implemented in the clearly defined territory (functional zone) and should be designed to achieve specific goals of the territory, that is, an appropriate combination of actions/measures must be selected after having assessed gaps and potential of the territory development in the strategy of fund investment. The essential task: identification of real demand and funding gaps by selecting a correct form and intensity of funding.

Effects of a longer period:

- Quotas for regions and drawing up of lists of regional project are no longer required (all functions of selection are integrated in the planning process); regional selection of projects will become significantly simpler (procedures will become up to half a year shorter) and the demand "to assimilate" allocated resources will be eliminated.
- "A bottle neck" in the system of regional planning will be removed – namely abilities of local partners and quality of projects.
- Investment of the Local development fund enables corrections (management) of municipal debt limits by managing them in a centralized manner for the most indebted municipalities, by making it possible to implement the projects of local development.
- It becomes possible to divert assignations allocated for municipality for various compensations, to promote a better collection and more targeted use of local taxes and levies.

- a system of tax motivation can be applied in due course. But as long as real opportunities of investment attraction are not assessed properly and a full-pledged set of tools and a system of incentives are not developed, redistribution of tax income may have an adverse effect, that is, it may demotivate municipalities economical situation of which depends on objective (mostly demographic and geographical) reasons.

2.4 Circular and Sharing Economy Strategies for Social Clustering

Circular and sharing economy is increasingly being discussed in order to ensure sustainability of public infrastructure and development of social clustering. As the Benjamin Franklin says, "Tell me and I forget. Teach me and I remember. Involve me and I learn." It can then be imagined that it would take time for the enabling schools for community based education and for community actors participate to raise a school. We cannot rely on local or state governments to take on the challenges we face locally. We need to empower the school so that it can facilitate the social community of clusters and encourage youth and society to face the complex challenges of our community needs and welfares.

We believe that the answer to the real transformation of education/school is a strong, authentic community connection and action. When families, community groups, businesses and schools come together to support learning, young people achieve more at school, spend longer in school and enjoy the experience.

The conception of circulate economy was developed to achieve sustainable growth in the context of an ever-growing demand for resources and pressure on the environment because of enormous rate of production and consumption. So far, economics was mainly based on the model "take-produce-throw away," that is, a linear model whereby each product inevitably expires. Valuable materials are used for food production, construction of infrastructure and houses, production of consumer goods and energy supply. These products after their use or when they become redundant are thrown away as waste. However, as a number of population and property grows, demand for scarce resources is growing more than ever and this leads to environmental degradation. This results in higher prices of metals and minerals, fossil fuel, feed and food, clean water and fertile land. Sharing economy (or collaborative economy, sharing economy, collaborative consumption, peer economy, demand economy) is a socioeconomic system based on sharing of human, physical and intellectual resources[1]. It is a new way to offer and consume products and services in special online platforms where you can find anything: accommodation for travelers, transport, tourism

1 Sharing economy generating billions of money – both danger and opportunities for the regular business. 2016. Internet access: https://euroblogas.lt/2016/06/07/milijardine-dalinimosi-ekonomika-ir-pavojus-iprastam-verslui-ir-galimybes/

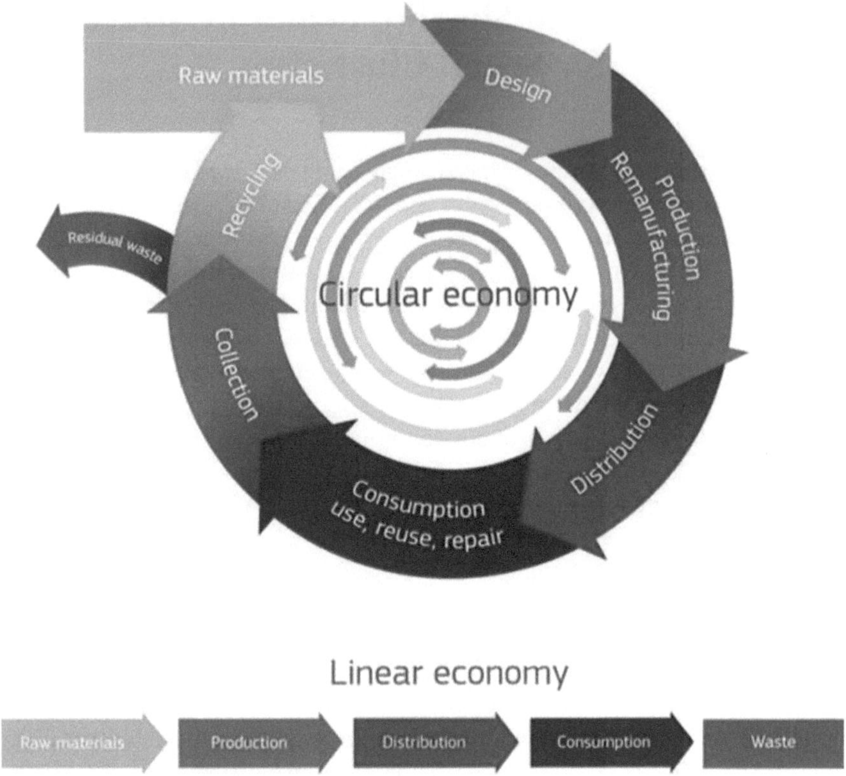

Fig. 24: Circular economy. Circular Economy/Sustainable Park Development. Internet access: http://gbbinc.com/services/circular-economy-sustainable-park-development

and catering services, or banquet/party service, home maintenance works, car rent. Sharing economy generated 28 billion euros of revenue in the member states of the EU in 2015. Revenue of its five biggest sectors (short-term rent, transport, home maintenance and repair services and finances) has nearly doubles over one year and continues to grow rapidly.

"Wall Street Daily" calculated that sharing economy merely in the USA created 60 thousand jobs and attracted 15 billion dollars. "Time" included sharing economy to the list of TOP10 ideas that will change the world. "PwC" forecasts that five sectors of sharing economy will earn 335 billion dollars of revenue until 2025. Public consultation and public opinion polls show high consumer interest in sharing economy. A recent Eurobarometer survey revealed that every third respondents has already used one or several of the platforms offered and that they would like to provide various services. People appreciate then services offered in this way are more accessible and cheaper compared to traditional services and payment can be made in the form of own services provided instead of money. Perhaps the biggest disadvantages of sharing economy are: suspense who should assume responsibility in event of problems, consumers do not trust sellers or service providers and online payment. LRV program (p. 74.2) continues the previous initiatives to give more responsibility to regions while implementing regional policy (consideration of suggestions of municipalities regarding projects of social and economic development of the region, selection of projects and submission of conclusions to institutions which grant funding), we will authorize regional development councils to make decisions regarding all issues of regional policy planning and implementation which can be addressed more effectively in the level of county rather than in municipal or state level. These councils which unite representatives of the residents of the county will have authorities to make decisions regarding the issues of transport, investment, environmental protection, health protection, social security, occupational training, culture and other which go beyond the boundaries of one municipality, the right to make decisions regarding allocation of investment of the EU funds, will be properly funded in order they could perform their functions. The Republic of Lithuania promotes compactness – after having assessed the ratio between input and benefit work, entertainment, places of science, sport and leisure time spending should be as closer to home as possible or should be easily accessible by public transport.

Circular economy (Fig. 24) aspects are directly connected to the Sustainable Development Goals of the 2030 Agenda proposed by the United Nations Member States (Fig. 25). There are direct links to Goals 7 (Affordable and Clean Energy), 8 (Decent Work and Economic Growth), 9 (Industry, Innovation and Infrastructure), 12 (Responsible Consumption and Production), 13 (Climate Action), and 15 (Life on Land) as well as indirect links to some of the other goals (Fig. 26).

Social and health protection infrastructure and quality of services in rural areas shall be improved. The basic quality health and social security services of the primary level provided at least in the municipal centers or, when necessary and if possible, in the larger municipal settlements must be easily accessible to the residents. Students who study the modules will understand the tasks of social clustering – to offer a higher quality and a sufficient accessibility abstaining from increase of input. Decisions are offered by sharing

Fig. 25: The global goals for sustainable development.

economy: the general use of the infrastructure, a temporary use, public and private partnership, models of mixed services, involvement of the community in service provision, concentration of high quality public services and regional cooperation of municipalities, integration of public transport).

Authors of the modules and students who study them must pose particular tasks for themselves:

1. To search for answers and definitions of quality services (what is a quality service?) and accessibility standard (how long it takes to receive a service of appropriate level by the resident and how many residents should have access to it). This standard (and the methodology of input-benefit analysis) shall be consistently applied during organizing of activities of all (irrespective of subordination) institutions.
2. Try to change attitude of managers of infrastructure, that is, from competition (struggle for survival of institutions located in remote areas) to cooperation while striving for the same goal – quality and easily accessible services. To motivate and increase pride in own work.
3. To develop territorial way of investment, to use more effectively investment resources divided between sectoral ministries and institution maintenance costs (place-based approach), to use the principles of ring and sharing economies. Development of the local development funds (LDF), empowering of local activity groups, investment planning on the basis of functional zones should form the basis not only for investment in the fields of regional

specialization (break-through) but also for investment in optimal operation of systems of services (a new model of regional planning shall be applied).

4. To eliminate Government and institutional barrier and an impediment to progress or movement, particularly in the form of regulatory oversight. Too much regulation and related bureaucracy can bog down processes in endless paperwork and delays. Inter-institutional barriers – restrictions established by institutions in the secondary legislation (hygiene norms, construction technical regulations) can be clear and should protect institutions in the area.
5. To enhance competence of regional development councils by delegating responsibility for optimization of services and for organization of systems of regional significance to them (by avoiding duplication of costs – when redundant infrastructure and management costs are developed and sustained due to problems of communication between individual municipalities and ministries).
6. To ensure that participation of residents and civil groups in dealing with local public affairs would become effective and prompting the local government to improve its activity on a continuous basis and to make decisions which best meet the needs of the local community.

Preparation of accessibility of services and quality standard (by defining uniform standards of quality of services (management, technological) and the requirements of accessibility): expansion of index of public service consumers' satisfaction (quality standard) in order it would encompass the tools of input-benefit analysis (a value for a price) and tools of territorial accessibility (by identifying services required in the local, municipal, regional and state (several regions) level and a time of accessibility (on foot, by public transport).

Assistance for regions in preparation of the plans of actions of integration of intersectoral and interinstitutional public service infrastructure and human resources, by promoting placemaking in the regions (by invoking regional chambers as the centers of competences and involving specific solutions to the plans of regional development);

All institutions operating in the central level accept competences of regional policy as necessity because the region "orders" the budget of specific institution (through the trend of specialization and specific financing gaps). The state becomes a partner instead of "a donor" which funds various needs.

The subjects analyzed in the virtual environment must be oriented to intersectoral target groups, must promote cooperation and partnership. The purpose and tasks of the modules are formed at the beginning of the module,

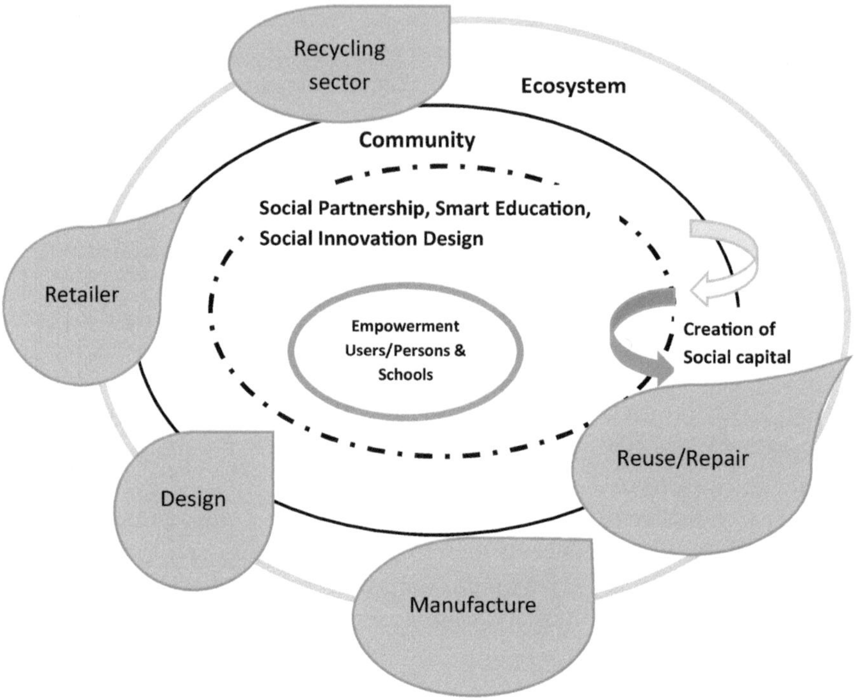

Fig. 26: Modeling of social clustering throw circle economy

the subjects (modules) and indicative preliminary sub-subjects should be constantly supplemented and reviewed by the authors of the modules in the course of the courses. During the course of training, the lecturers and the students shall revise competences which will be acquired by the participant of training who has completed the Program and training (learning) methods. The ways of evaluation of acquired competences are consistent with teamwork, cooperation and clustering demanding creative tasks.

The main literature designed for the participants of training must be accessible virtually (at least five sources). Additional literature must be indicated too.

Additional texts of training material (scanned material of monographs, textbooks, articles required for analysis of the subject and/or original texts supplementing material set out in PWP slides which is required for presentation of the subject, videos and photos) must be interactive. During the studies of the module, they should be supplemented with the best creative works provided by the students.

Picture 4: MA and BA students with member of Lithuanian Parliament Justas Džiugelis after discussions about inclusive education

Description of independent tasks, provided time of tasks performance, questions for self-test, practical tasks. A provided mechanism of tasks evaluation shall be presented for each subject of the module at its end: the final way of practice or a form of control work which may be: the project of social business, test, project, study plan, creative work, Flash mob, presentation of an idea based on creative industries and social business: video, text, project Padlet[2] platform – loaded to a virtual medium, etc.

Literature intended for the subject analyzed shall be specified at the end for analysis of the module. Such literature shall cover national and international legal acts, national and international theoretical and methodological literature, references to databases (all literature must be accessible virtually or a scanned material shall be presented in the training material of virtual material). The total volume of material of the training module cannot be less than four author quires. Training and methodological material must be prepared in Lithuanian and English. A text of the material must be edited properly, formatted (when necessary) and all material must be fully prepared for use during training. The final prepared text of the material should be absolutely ideal, no corrections or other works of perfection should be required. Training and methodological

2 Internet access: https://padlet.com/slaatto1/fkcgx5pzsa16

Picture 5: Morten Flaten Paulsen (Norway) together with Social Education Faculty after discussion about challenges of virtual learning in the Lithuanian University of Lithuanian Sciences

material must be submitted in electronic medium or by sending to agreed email address and one printed copy of the material shall be submitted. Text documents shall be presented in Word or equivalent format.

The requirements for training:

To prepare a scenario for videos of training (video material) to be used for presentation of training;

To present training during shooting (for preparation of video records). A number of presentations: at least 12 pcs., four presentations for each module (works of shooting will be performed by other executors selected by the customer);

To prepare and agree times of communication (at least 15 hours in total) for consultations of the participants;

To implement training using a virtual learning environment of the customer in the Moodle system:

- to provide information to the training participants;
- to consult the participants;

- to provide remarks, answers, comments in writing
- to carry out monitoring of the participants' learning, to notify and remind about outstanding tasks when the participants are late to perform them, to prepare and send out necessary notifications;
- to perform evaluation of tasks, to submit generalizations;
- to discuss together with the participants tasks completed;
- to perform all other actions directly related to training.

All required actions during the course of training should be performed in the Moodle system.

III Smart Education Strategies

3.1 Analyses of Success Case for Social Entrepreneurship Scenarios

In this monograph, we will compare family-friendly policies and smart education tradition and case studies in Lithuania and others countries, with specific reference to their policies supporting childbearing and childrearing. In general, terms, family-related policies, which are partly designed to increase fertility, can be classified into two types. One type purposes to reduce the economic burden associated with raising children. Gamification describes the principles of adaptation of the game in education, business and other processes that are relevant to the consumer experience and commitment to dominate in nontraditional contexts. Indiana University[1] has some of the longest experience in the field of education. Prof. Dr. T. C. Mason, the dean of Indiana University School of Education, together with other IU experts, hosted open lectures and international conference reports in Lithuanian Educational University several times.

Indiana University School of Education is one of the biggest educational schools in the USA, which focuses on educating teachers, future educational institution managers and administrators, as well as scholars and educational evaluators within the network of several units operating within IU. School of Education researchers are leading actors in Civic Education Networks and pioneers of integrated educational innovation in Secondary Schools. The long-term partnership between Indiana University and Lithuanian Educational University, now: Vytautas Magnus University and Lithuanian College of Democracy has encouraged us to create joint publications[2].

---

1 School of Education. Internet access: http://education.indiana.edu/

2 Kvieskienė, G., Masson, T. C. Integrating Civic Education and Social Work: The Social Educator Master's Degree Program at Vilnius Pedagogical University. In: Civic Education in Teacher Education, Volume 3, The Social Studies Development Center of Indiana University in Association with Civitas: An International Civic Education Exchange Program. Edited by Gregory E. Hamot, John J. Patrick and Robert S. Leming: Indiana University Social Studies Development Center, USA, 2004. ISBN 0-941339-28-9, p. 189–215; (2). Kvieskienė G., Metcalf, K., Nixson, W.A, Patrick, J.J., Vontz, T.S. Implementation of Project Citizen in Indiana, Latvia and Lithuania, in Project Citizen in Indiana, Latvia, and Lithuania, USA: ERIC, 2000, p. 103–125. ISBN 0-941337-25-4; (3). Kvieskienė G., Žadeikaitė L., Patrick J. J. Constitutional Democracy in Lithuania//Civic Education in Teacher Education. Volume 4. The Social Studies Development Center of Indiana University in Association with

Researchers analyze school clustering that in turn offers educators a common sense, and school as a whole approach to education innovation and staff development. This model focuses on what teachers can achieve by creating a networking culture and enhancing every teachers' capabilities.[3] Indianapolis' vision for Smart Cities is to support a connected, healthy, resilient, inclusive and competitive city. Indiana University has utilized cluster hiring to elevate their institutional profile. Researchers agree that the lack of research on clustering effectiveness reduces possibilities for innovation and collaboration.[4] Moreover, Indiana University Library is one of the highest ranked in the US[5]making it one of the most appealing research environments. Post-modern culture is a network culture. For us, fish net symbolizes connections, especially between NGOs and the public and private sectors. Fish net symbolizes sharing, teamwork, social partnership and social clustering culture. The network calls for our connections to be useful: to catch fish and seafood – in other words, to develop a senses of 'welfare' that is of a new quality. The possible conflicts between individualization and cooperation, and between localization and globalization are challenges for the theory and practice of contemporary social work and social pedagogy. Fostering an individual identity while accepting differences is a real challenge that networks can help to meet. Our experience has been gained through conflict mediation and creating network organizations to consolidate the socialeducational professions and help to resolve social problems at local, national and global levels. Building trusting and authentic relationships is very important for modern social work and social pedagogy. Through relationships we can solve social, educational problems, show society how we can manage

---

Civitas: An International Civic Education Exchange Program. Edited by Gregory E. Hamot, John J. Patrick and Gregory E. Hamot: Indiana University Social Studies Development Center. 2005. 180–217 p. ISBN 0-9773916-0-4

3 Marcia Gentry, Ph.D. WITH Kristina Ayers Paul, Ph.D., Jason McIntosh, C. Matthew Fugate, Ph.D., & Enyi Jen. (2014). Total School Cluster Grouping & Differentiation. Internet access: https://www.education.purdue.edu/tscg/documents/Total-School-Cluster-Text.pdf

4 Susana M. Muñoz, Vincent Basile, Jessica Gonzalez, Daniel Irmingham, Antonette Aragon, Louise Jennings, & Gene. (2017). Gloeckner (Counter) Narratives and Complexities: Critical Perspectives from a University Cluster Hire Focused on Diversity, Equity, and Inclusion. Internet Access: https://lib.dr.iastate.edu/cgi/viewcontent.cgi?article=1139&context=jctp

5 Indiana University—Bloomington. Internet Access: https://www.usnews.com/best-graduate-schools/top-library-information-science-programs/indiana-university-bloomington-151351

them, become role models of social partnership and social innovation. Each of us is a unique person and only after an assessment of this uniqueness can we find out what benefits can be received when pool our collective talents.[6]

Gamification[7] is a new way of thoughts how game elements can improve systems and engage users; yet continuing the body communication, environment in all this area. Educational games and simulations complexity allows not only the integration of multi-disciplinary skills and competencies, but also strengthens the motivation, increase student involvement in education (learning) and the educational process, provides the possibility of using nontraditional spaces and development of persons to know yourself changing circumstances. It is not substituting previous work but extends the ideas into creativity and innovative ways of thinking about system development that actually considers the user as a crucial component that needs to be rewarded and motivated. Today, the whole of society is in search of active inclusion and motivation. Reserve, faced with the huge life-long learning and smart educational challenges. Gamification or game elements in the play as well as support the development of the script, give more opportunities to help educators select the most appropriate scenario to solve complex new knowledge and skills in mastering problems. Families with disabled children in different European countries by identifying this problem we initiated interview of Representatives locally and globally from arena Education Researches, innovative private, public and civic sectors.

We send 250 questionnaires for Select Representatives. In same way we Acknowledge 165 Questionnaire, 49 from Civic Society, 101 from Expert and have Focus group discussion with 21 national and international expert's education project leaders. In focus group, we ask the question: which indicators in project managing are the most important for Sustainable and smart Education? Experts choose:

Educational experts chosen from Lithuania, Finland, Netherlands, Spain, United Kingdom, Belarusian, Norwegian, USA and we analyze what kind of experience from smart education interfaces with social capital they a familiar or involved in educational institutions, where they are involved. Our research focuses not only on smart education, but also describes general experiences on social partnership and social clustering and possible recommendations for improving the situation while adapting the good practice models of Norwegian,

6 https://socialworkin40objects.com/2016/07/17/mobile-garden/

7 Gamification of learning. Internet Access: https://www.linkedin.com/learning/gamification-of-learning?trk=slideshare_sv_learning

Tab. 7: Sustainable and Smart Education? Experts Choose:[a]

Indicator	Experiences, Good practices
1) **Systematic approach**;	Team working, (PPP + Schools, NGO, Universities, Social Services, Business, etc.), Networking, Clustering, database, case study, second chance, alternative programs);
2) Social-**Emotional intelligence:** system, crisis intervention, consulting, guidance, social mediation, mentoring.	Sustainability, environment, climate, creativity, emotional, natural setting =IQ (better teachers, quality of school, green school, social partnership);
2) **Inclusive aspect**: skilled and innovative based social researches,	Social pedagogues, youth coaches, one contact system, crisis intervention, consulting, guidance, social mediation, mentoring.

[a] Giedrė Kvieskienė, Eglė Celiešienė, The Development of Indicators For Sustainable Education: Methodological and Conceptual Approaches// THE DEVELOPMENT OF INDICATORS FOR . . . Internet Access: www.bm.vgtu.lt/index.php/bm/bm_2014/paper/.../560

Denmark, the Netherlands and United Kingdom through interviews with 18 experts.

The paper relies on four case studies of NGOs that have long-term social partnership with education institutions and have social clustering practices. In our research, we find three types of social clustering experiences with leading educational institutions. There, according to respondents, they have more positive experience, without possibility to be bullied or left behind academically. They also state that home schooling or family friendly education provides even more positive results.

Every local authority and institutions across the Lithuania are responsible for education institution environment and can manage create own portfolio of community activities and innovations, yet there is no structure or funding which ensures a proportion of these are community or innovation based. In a bid to change this Practice has been working tirelessly with a number of great organizations and individuals to create SE&SC (Smart Education for Social Clustering) plans, which the process of designing smart education scenarios much easier. There are not easy to be sure that we involve everyone school and community stakeholders, especially the child's and parents you are mainly adapting these scenarios for. Our experts mention, that there are important to include questions to consider four habits asking children, like:

Tab. 8: Four habits for Smart Education and Social Clustering Skills

1. Investigation in Making Acquaintance and Beginning: Holistic Learning	**Set common Rules Together: Trust, Development of Social Capital**
2. Together, discuss perspectives: Personalization: Orientation to all Process Participants: Empowerment	Listen of Students expectations, discussed supplement their rules, set goals Social
3. Demonstrate empathy: Relationship	Communication and Positive Socialization and Experience
4. Invest in recommendations and generalizations: for wellbeing and Happiness	Involve 3M/D Principle: multifunctional, multicriterial, multisectoral ©1996, 1999, 2003 The Permanente Medical Group, Inc. Physician Education & Development Revised April 2003 in partnership with the Kaiser Permanente Institute for Culturally Competent Care

Making decisions, which are important to let all community members (children's, parent, stakeholders, partners (make their own decisions. If the any of circle member has trouble communicating, using the SE&SC system will help. This will allow you to give everyone a range of choices, but the child will make their own decisions.

When you are planning, start clustering process: think about different ways of communicating. For example, with some children, parents or partners respond better to write official agreements, arrange meetings, workshops or open meetings or open working spaces, non-formal education projects, exhibitions, other events. It is important that you plan clustering process appropriately and that you keep the process clear, structured, tidy and make necessary designs:

- involve all actors and stakeholders in regular open meetings in which can participate any, who have ideas and proposals;
- Make presenters and activate social networking about achievements and share it with your circle of partnership;
- Sign formal agreements and make them public for your communities;
- Look for common goals and mutual benefits.

As confirmed by Indiana University researchers and our research, upon using the four-step positive communication model, social educators, other education practitioners, child caregivers and licensed nurses report significant changes in emotion in positive parenting and social educator consultation with parents

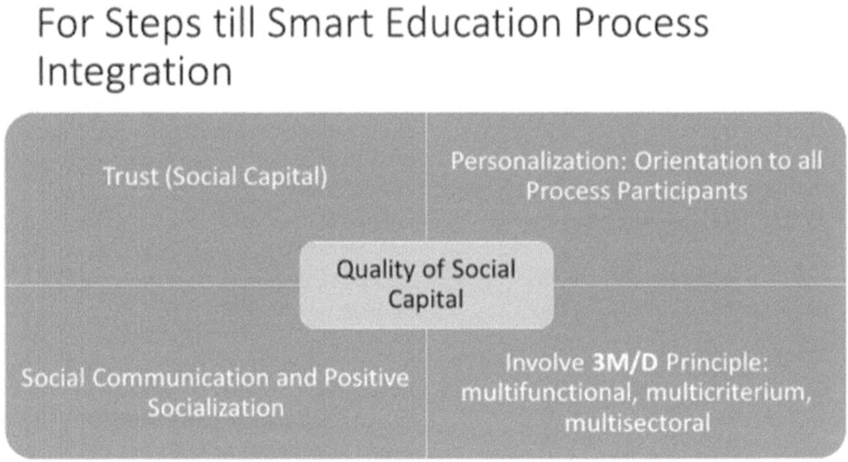

Fig. 27: For steps till smart education process

in three areas: communication skills, relationships and trust. Previous Studies by Frankel and other researchers have shown that this four-habit positive communication model has positive long-term effects on both the physician and the patient. This model is widely used in the United States and other countries to train doctors and nurses.

Dr. Frankel is a professor of medicine at Indiana University, and director of the Maryan Margaret Walther Palliative Care Research and Education Program at Indiana University. Each step proposed by these researchers contributes to a four-habit model of positive communication that contributes to better integration of childcare, social communication, and children with disabilities. Responding to the needs of children, their parents, caregivers, we will better understand the needs and social responses of child and support professionals that contribute to better educational and therapeutic outcomes, and our research confirms that twenty-three skills can be complemented by additional short- and long-term interventions described by our students. Parental involvement and empathy, increases parental and caregiver engagement and positive socialization scenarios in all steps. From case studies, we can conclude that social and education professionals are able to use several new social communication scenarios in emotionally charged situations, and are an effective way of exploring the invaluable relationship between health, social care, nursing and education institutions professionals and parents clustering, when addicts or other risk children are vulnerable and professionals are out of touch with parents to incorporate their knowledge into all day child education and care plans. Qualitative research by researchers at Indiana University to determine parents' and caregivers' perceptions and experiences with the nurse. General communication theory was the result of this study, which included interviewing, listening, explaining, speaking, understanding, and role negotiating to achieve scenarios and mutual understanding for a childcare plan and positive communication. Educational support professionals and, above all, social educators need to understand parents' understanding of working with families to optimize the care they provide (see Tab. 7).

We try this strategy the project "Play for children with disability: case studies"[8]together with partners from Sweden, Finland and Lithuania. Such multifunctional and multisectorial sensory experiences are very important for all children and their parents. The four steps of positive communication give us opportunities to experience different sights, sounds, smells, tastes, and textures. Many children with Additional Support Needs do not have the opportunity to explore each of these senses for various reasons. Play for children with disability may be a topic for various professions: preschool teachers, social pedagogues, special needs educators, leisure time educators, teachers in physical education, occupational therapists, psychologists, youth workers and may be treated in the study of early interventions, language and speech therapy, special educational needs, inclusive education, developmental psychology, psychomotor development, neonatology, disability studies, assistive technologies, etc.

The role of play in the research is rudimentary, and it is seen as additional leisure activity, with little to no focus on it in children's schedule. The case studies in this paper offers a viewpoint on special needs education in Lithuania, which emphasizes not inclusion but a certain level of marginalization (sending children into special care home instead on working individually with them in local communities). To solve this, a methodological tool "Individualizing educational programs for students with SEN "was created. It's a set of recommendations for

8 Westling Allodi, M. Lundqvist, J., Lundström, E., Roll-Pettersson, L. Rosendahl, J. Siljehag, E. & Zakirova Engstrand. R. Play for children with disability: the users' needs in the Swedish context //Westling Allodi, M. Lundqvist, J., Lundström, E., Roll-Pettersson, L. Rosendahl, J. Siljehag, E., Zakirova Engstrand. R., Suhonen, Eira; Takala, M., Alijoki, A. and Viljamaa, E., Celiešiene., E., Kvieskienė, G., Kaušylienė, A., Trukanė, I. Play for children with disability: case studies. Case studies from Sweden, Finland and Lithuania. Internet Access: https://www.degruyter.com/viewbooktoc/product/507228?rskey=BjKH9q&result=4

Tab. 9: Four Habits for Smart Education Skills

Develop Habits	Develop Skills
1. Invest in Starting and Setting Rules Build a relationship of trust with children	Concern over common problems; (1); Concern over common problems; let the children know what they can expect in the future (2). Understanding and maintaining individuality (3), non-formal environment (4), non-formal leadership and partnership (5).
2. Use and demonstrate the priority and perspective of the child's (client's) interests	Find out the child's ideas, needs (6); Set a specific query or goal for the client patient (7); Investigate the impact on the client's (child's) biological family, relatives (8). The natural environment, nature and being comfortable (9); Informal collaboration, simple, natural materials (10),
3. Demonstrate empathy	Demonstrate empathy Be open to the client's (child's) emotions (11); make your statements sincerely (12); convey sensitive concerns at the nonverbal level (pauses, touches, facial expressions (13); Friendly environment, lots of visual details and opportunity to participate (hands, eyes, heart) (14); Few cold colors; is "where to help myself" (purposeful activity) (15);
4. Summarize the suggestions at the end of the consultation.	Present the positive socialization scenario (16), the complex assistant plan (17), explain the proposed measures and activities, counseling (18); summarize the immediate plan (19) and the positive communication scenario (20). Lots of collaborators, supporters; informal connections (21); networking (22), social clustering (23).

social pedagogues and teachers, who work with children with different needs and abilities. Sensory experiences are very important for all children. Play for children with disability may be a topic for various professions: preschool teachers, social pedagogues, special needs educators, leisure time educators, teachers in physical education, occupational therapists, psychologists, youth workers and may be treated in the study of early interventions, language and speech therapy, special educational needs, inclusive education, developmental psychology, psychomotor development, neonatology, disability studies, assistive technologies, etc.[9]

Likewise, maternity leave and parental leave programs were created to protect employees, promote gender equity, or achieve the effective use of the labor force. The construction of a family-friendly society is nowadays an important policy target for industrialized countries with the blew-replacement level of fertility. The most family used to live three generations together. Adults other than children's parents in a community used to attend and educate to the children instead of their parents as well. There was a strong connection with local people. Therefore, children naturally experienced to play with others in community of different ages, take care younger children, and learn from older children. In this way, children were taken care not only by their parents but also by locals more than today. Lithuania is currently developing a network of children day care centers, which provide services for children from families at risk of social exclusion. These centers do not only provide various after school activities to develop the social skills of children, but also psychological, pedagogical and legal assistance to their parents. The establishment of these centers is seen as an important contribution to diminishing the number of socially dysfunctional families. Family therapy[10] in Lithuanian practice traditionally addresses only for the social disadvantages families for therapy and consultation with participation NGO or professionals institutions for better their relationships, include important persons in their lives and empowering other social networks and stakeholders Family therapy and systemic practice in Lithuania are poor, because, we can use only 8 hours for that process. But we understood this process as a systemic arena; „some of the commonly shared goals of family therapy might be for example: improvement of family functioning on different levels, enhancement of mutual understanding and emotional support among family members, development of coping skills and problem-solving strategies in various life situations. "Systemic Family Counseling (SFC) programs are about

9 Celiešiene., E., Kvieskienė, G., Kaušylienė, A., Trukanė (2019). Play for children with disabilities: the users' needs in the Lithuanian context //Mara Westling Allodi and Tamara Zappaterra (Eds.). Play for children with disability: case studies. Case studies from Sweden, Finland and Lithuania. Internet Access: content.sciendo.com

10 What is family therapy? Internet Access: http://www.europeanfamilytherapy.eu/what-is-family-therapy/

helping parents to better understand their child's needs and behavior including smart socialization methods for happy and healthy life".[11] Positive Parenting through Smart Socialization (PPSS) program is part of Systemic Family Counseling (SFC) module and designed to improve parenting and children social and creativity skills (Fig. 1) (Kvieskienė & Celiešienė, 2016). Integrated family support, emotional therapy and integrated practice are heterogeneous areas. One goals of family therapy or complex assistance can be, for example, improving the functioning of the family at various levels; Strengthening mutual understanding and emotional support from family members; Ability to solve problems and develop problem-solving strategies in different life situations. Family Assistance is designed to help parents better understand their child's needs and behavior, including positive socialization scenarios for a happy and healthy life. The sample is comprised of Vilnius teachers working according the preschool curriculum and social educators, social workers, social pedagogues, social educators and psychologists – those working with children, working in families and therefore particularly concerned with child well-being. Following the presentation of the Four-habit model of Positive Communication by the researchers, it was asked to highlight the steps to be taken to strengthen the competencies of teachers and childcare professionals. The analysis of child welfare reveals that Lithuanian laws are not harmonized with the UN documents, the Convention on the Rights of the Child, there is no monitoring and information collection and evaluation system. (Tab. 7).

Before modeling new services, modernizing educational, social, multifunctional, cultural, health and other public institutions and strengthening partnership of the public and private sectors we must answer to the questions how we can enable public institutions and spaces to become alive space of creative and social industries that is able to initiate new services by smartly using infrastructure and environment. One of the possible principles trying to ensure a constant interaction and synergy of different sectors between the public, private, NGO, science sectors and marketization of services and new products is the 3 D model of smart education (multifunctional, multisectorial and multi-criteria), that is, to seek for the maximum decentralization of services by supporting initiatives of civil organizations and private initiatives. Success cases of smart education can be found in Lithuania, Norwegian, Belgium, Holland, Denmark,

11 What are evidence-based programmes? Internet Access: https://www.actionforchildren.org.uk/what-we-do/ services-for professionals/evidence-based-programmes/

Canada, the United Kingdom, Ireland and other European countries and the countries of the rest of the world.[12] Their analysis confirms the fact that multi-functional, multi-criteria and multisectorial design of institutions and organizations helps neutralize segregation and stigmatization of social groups (ethnic, religious, social, age-based), increases inclusion, creativity and innovation. Multifunctional design, social communication and partnership facilitate marketing of services and goods and increase growth of intersect oral social capital.

The four-step positive communication scenario is used to address the complex issues that are most often associated with empowering children, their parents, and solving complex issues. Therefore, "it is necessary to strengthen children's rights and adopt a child-centered approach /. . ./to formulate clear evidence-based policy goals /. . ./to promote integrated and coordinated policy-making and implementation /. . ./to include children among key social policy themes" (Frazer, Marlier 2012: 21, 22).

Therefore, in order for education aid to be effective, it is essential that funding is balanced between education, health and social security.

Smart Education is one of the most receptive in terms of creative and social industries because it can be most easily implemented through a positive socialization, experiential education, scenarios of experience, adventures, dreams, playful and smart education that are characterized by four main features: (Kvieskienė, 2005, Kvieskienė, Kvieska 2012, Putnam 2015):

- Commitment – to assume responsibility and to be open to changes;
- Objection – to focus on tensions between the actual and desired practice;
- Conflicts – to use energy for positive solutions;
- Challenges – to change of typical reactions, believes and attitudes in non-dangerous way.

Innovation and education, entrepreneurship and social clustering are the prime ways that growth and inclusion in Europe can be supported, says the White Paper, which puts forward key recommendations for action: from supporting public-private partnerships for technology diffusion, investing in smart transport infrastructure, supporting schemes for migrant entrepreneurs, a better strategy for teaching advanced digital skills as well as better access to venture capital and nontraditional sources of financing for small businesses.[13] Working

12 A map of social enterprises and their eco-systems in Europe Executive Summary. Internet Access: www.socent.ie/.../a-map-of-social-enterprises-and-their-eco-systems-in- europe-executive-summary/

13 Internet Access: http://www.prosperity.com/feed/regional-analysis-europe

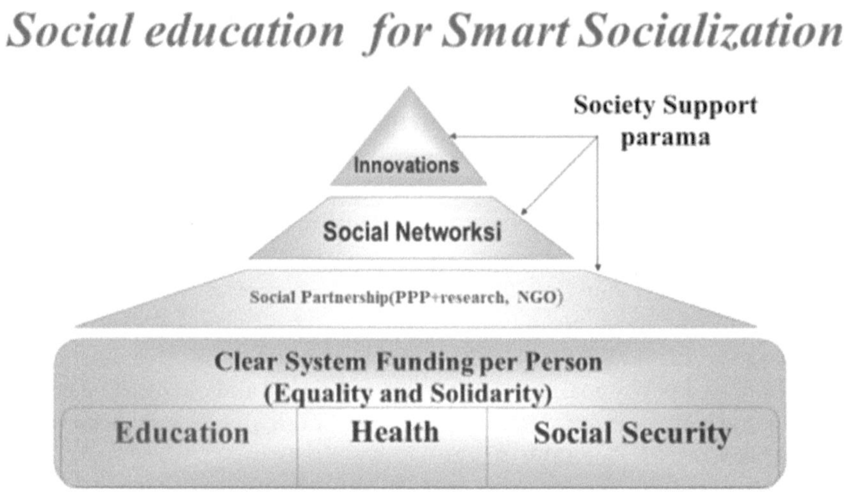

Fig. 28: Social education for smart socialization

strategically with the social cluster's initiation in small communities can help to prioritize the efforts.

We have characterized a conceptualizations of Smart Education (see Fig. 5) as open, associative, while the model we have defined, based on Koper's (2014) work, is proposed to be more complete. We have wanted to develop a model that can be leading by education institution and integrate new smart innovations without having to introduce new system elements.

To test the completeness of the model we contrast it with the framework of smart learning introduced in a recent book on smart education in smart communities and smart cities. Will the four types of support technologies for smart learning identified in their framework easily be integrated in the SLE model? Figure 5 describes the smart learning framework of Liu et al. (2017). The model positions the learner in the center and consists of four levels (learning experience, support technologies, learning scenarios and basic principles of teaching and learning). Four support technologies for Smart Education are identified. Awareness and adaptive technologies, artificial intelligence, sensors, and auto deduction are mentioned with applications that identify type of learning situation, provide diagnosis of learners' problems, personalized learning resources, social matching, and suggestions about learning activities.

1.	Smart Education for sustainable environment. **Identification.**
2.	Education Competitiveness and Social Differentiation Index **Alternatives and Promotion of Diversity.**
3.	Integral Sustainable Prerequisites Educational Index: Impact and **Autonomy Insight.**
4.	Modeling of Positive Socialization: Active citizenship and inclusion. **Positive Socialization Insight.**

Fig. 29: Smart education for sustainable environment

The conception of social business[14] is the phenomenon that is the closest to the social clustering and smart education. Social business is a rapidly progressing field in Lithuania. In this year, the forum of social business has grown from search for ideas in 2012 to the line of LRV program, real investment of the EU and the national investment, the Law on Social business preparation of which now is in progress.

Each Generation has a distinctive "generation consciousness," which is dependent on the pace of social change (Erstad 2010).

The features of smart learning as a smart perception, strong interactions and high immersion earn appeal to digital learners learning needs. (Huang at all 2012) However, traditional classroom environment is not beneficial to the development of smart learning, and the learners have relative low satisfaction from learning experience in current classroom environment. The performance level required from the smart classroom is far for more complex that the one requirement in traditional classroom teaching (Ringstaff et all 1991) Smart classroom is a typical materialization of smart learning environment and is the high end form of network classroom, where "Intelligence" involves five dimensions: Showing, Manageable, Accessibility, Real time interactivity and

14 A guide of social business. Internet Access: https://www.business2community.com/social-business/start-social-business-guide-01651630

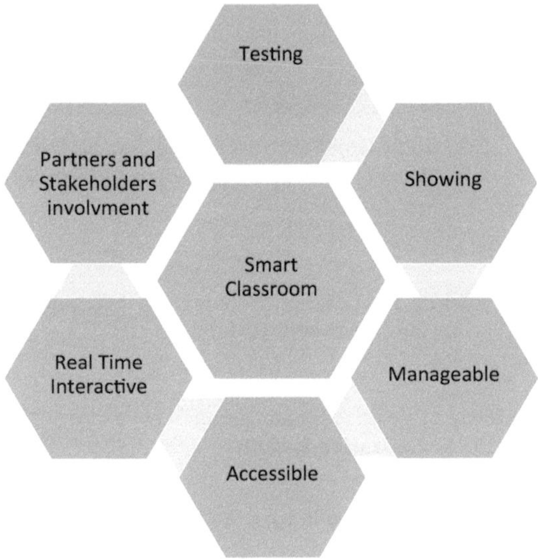

Fig. 30: Smart classroom model

testing, abbreviated as SMART Conceptual model, showing as (Fig Huang et al., 2012). Each generation has a distinctive "generation consciousness," which is dependent on the pace of social change (Erstad 2010).

Demand for integrated social and education services, an action focused on family and community groups approaches changes and real needs and social business startup training is huge. The studies reveal the scissors of parents' expectations for the future in relation to the child and the existing situation: many parents imagine that their child will be become an independent, self-confident adult able to communicate. Now they cannot identify their children as such. After having analyzed the methods of children's education applied it becomes evident that many parents imagine that they positive methods of education in education of their children. However, control situation showed that perception of parents and practical abilities do not match. Many parents also still tend to punish their children during the process of education by criticizing and moralizing them, even by applying forms of physical punishment. Only some of them manage to allow their children to feel natural consequences of their behaviors. S. Burvytė (2012) claims that many parents gain knowledge and skills on the issues of children 's education from experience of their parents and only some of them tend to seek advice of professionals, to contribute themselves and

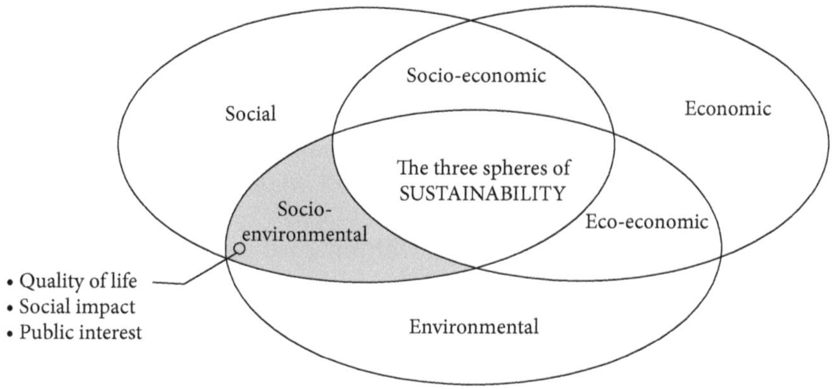

Fig. 31: The three spheres of sustainability

stimulate their children to be active in the local communities. Results showed that first of all a concentrated activity of public, private institutions, civil society in sake of welfare of the society and leadership of a social educator will help for humanization of the proximate environments and development of innovations. In order to help control these arising interferences we should adapt the following five principles:

1. **Personalization**, creation of personalized, informal environment (creation of family friendly environment in institutions, development of open, multifunctional spaces, informal environment);
2. **Originality,** creativity, innovation, flexibility, openness, team projects and solutions (virtual training, tolerance, recognition of different skills);
3. **Volunteering,** sociality (good works of societies, interaction with the community movements, NGO);
4. **Development programs** (safe neighborhood, partnerships; networking, social clustering);
5. **Social partnership of public, private, NGO sectors** (integration and development of philanthropy, social, responsible business, social and sharing economy).

When you start to create new social cluster, you can enable schools for smart education and involve partners and stakeholders from social, economic and environment spheres (see Fig. 31), which are around the school. Quality of life, Social impact and Public Interest are mainly concentrate in Socio-environment sphere.

The volatility and complexity of the evolving social and challenges of education environment place increasing smart education demands on our teachers and new fields of responsibilities for community's stakeholders. To thrive under these conditions, teacher and school management require a broader set of competencies, higher levels of proficiency in them, and a greater ability to rapidly learn new material to confront novel challenges. In other words, Teacher and school's principal personnel must develop an ever-expanding set of sophisticated, agile knowledge and skills – with minimal significantly increasing training and education time or costs. Our monograph research as the other our cited publications point toward a shared vision, which includes the concept of a modern, smart "learning ecosystem" comprising interconnected learning opportunities, supported by technology, driven by data, and integrated with other talent management capabilities.

This envisioned "continuum of learning" includes features such as the following:

- Continuous: Career oriented-long, continuous learning replaces the status quo's stovepipe, episodic learning
- Blended: Formal training and education, learning by doing, project oriented, just-in-time support, and informal learning are integrated
- Enterprise and community need based focused: Training, education, and talent management are considered in concert, holistically
- Diverse, universal design based: Disparate learning technologies and methods are stitched together into a cohesive ecosystem
- Learner centric: Learning adapts to individual and team needs work based, contexts, and characteristics
- Data driven: Learner data from across many sources are aggregated and analyzed to drive decisions
- Competency based: Competency frameworks support assessment and smart education scenarios, research and case studies, guides developmental trajectories
- On Demand: Personalization and individualization. Modular training and education can be delivered at the point of need
- Cloud Based: Software services and network-based repositories support flexibility and discoverability in smart and sustainable learning environment.

This strategy called Total Learning Architecture (TLA) is a collection of specifications for accessing and making use of learner-related data. We often in smart education topics use the term "Big Data." TLA seeks to leverage that

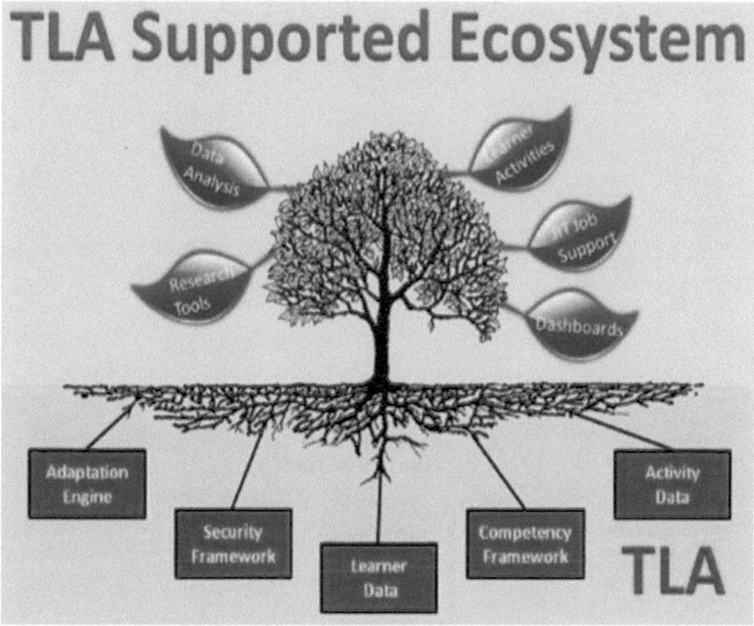

Fig. 32: Total learning architecture

data to create more flexible and effective learning environments. A major part of TLA would be the sharing of data across networks and the portability of data. This is similar to what has been proposed for individual medical records the last several years. There are a lot of the same benefits. It also has some of the same concerns (See Fig. 32).

Ethical, civil, economic, or simply social responsibility is closely related to activities of the local communities. However, their development is optional and they are not governed by the Law on Education or by the documents of schools. Therefore, an aspect of values internalization cannot be left on the side when developing a social clustering and partnership for a sustainable development. An integrated attitude is necessary to this end. Philanthropy and empathy develop intersectional confidence and help develop good citizens. Therefore, this is important for not only the citizens of general educations but also for communities and the society. Well-known pyramid of social responsibility developed by Carroll is the first attempt to perceive components and cohesions

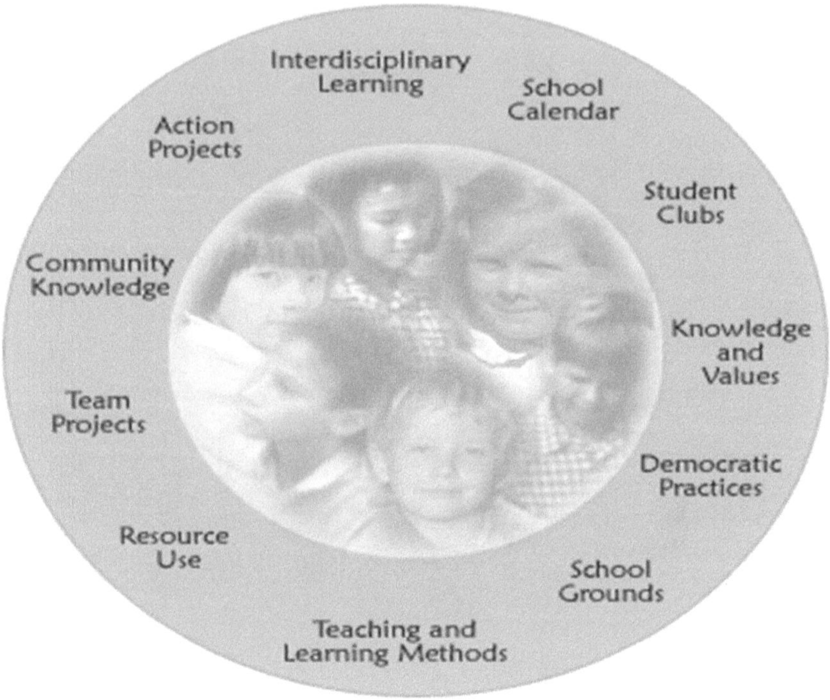

Fig. 33: Learning for sustainable development

of social responsibility. Based on Alban and Kelabi (2014) responsibility must be validated as an imperative of all economic activities[15].

In order to define and distinguish methodological accesses or attitudes presented (paradigms) which help to analyze in more detail the results presented and good practice we rely on attitudes of positivism, humanism and systemic attitude and development of social industries, social communication in different sectors. Together with other innovative technologies, such as social partnership (Kvieskienė, Kvieska, 2012), positive socialization (Kvieskienė, 2007), community oriented sustainable development (Kvieskienė, Bardauskienė, 2014), learning by doing (Targamadzė, 2014, Bessen, 2015), sustainable development of social clustering should become a measure of social exclusion and systemic changes.

15 Analysis and critical evaluation of the ethical standards of the UK's No.1 Sports Retailer. Internet Access: https://ocinternationalmanagement.wordpress.com/

Methodological access of humanism (paradigm) mainly focuses on the natural phenomenon that reveals freedom of an individual and emphasizes systemic, holistic, Ideograph, contextual methodology. Humanism is also described as use of individual or social experiences and proficiency gained. Systemic methodological access (paradigm) involves an external "objective" model and an internal "subjective" experience of an individual. This attitude explains what a place intercultural transformation is and why it is in that place. Therefore, the main trends, such as "stress, adaptation, a growing dynamic" encourage an individual to create his intercultural "capital." According to W. B. Gudykunst (2005), an attitude chosen and criteria of environment assessment essentially determine how intercultural communication that, besides cultures, also involves different social and age groups is carried out. A content of systematic welfare scale is described fairly in detail by sufficiency of resources disposes, which are necessary for a power of systematic capacity taking into account a constant flow of challenges (Dodge, Daly, Huyton, Sanders, 2012). Depending on the nature of the system, these challenges and resources can be psychological, social and physical. "The manual of transparency policy of state welfare implementation" that was prepared during the project for different public, private and NGOs sectors" attributes typical set points of individual state of capacity and individual codes of distribution of generation of necessary resources between the goals of welfare development.[16]

A sustainable development of social clustering can be identified as shift of ideas, thoughts, values and attitudes when representatives of different cultures or subcultures, age or social classes communicate and cooperate in different educational environments as well as partnership and a dialogue based on problem solving that changes during exchange of culture, attitudes, values and historical experience. However, more detailed describing of meaning of a content of social education depends on training subjects selected and a field of an attitude. Theories analyzed reveal that the same phenomena can be analyzed in different way, using different assessment criteria and organizing activities, using the theories of social education that highlight different aspects. Therefore, these aspects can be important in preparation when getting ready for integration of different training subjects.

In order to look deeper into "The Post – 2015 UN development agenda" of the UN it can be seen that both assurance of a global economic growth, increase of potential of individual regions or country's growth and solving of emerging

16 More Look at: The manual of transparency policy of state welfare implementation. Internet Access: www.socialclusters.lt

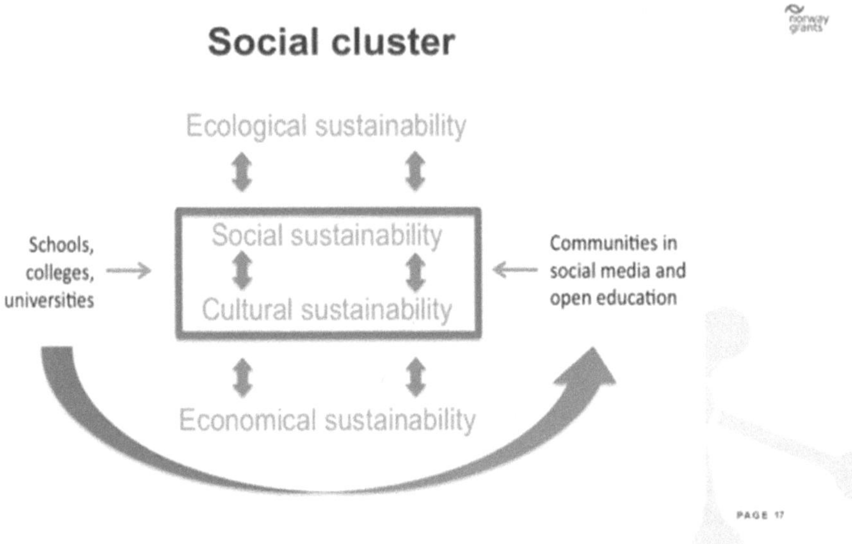

Fig. 34: Social cluster

global problems related to food, health, increase of ecological unsustainability can successful only using a purposefully developed cluster of knowledge, innovations and technologies (SKNIT). For example, we do not have alternative for purposeful development of scientific knowledge and space technologies when speaking about possibilities of global or space accidents. Future challenges and ways of assessment of their potential consequences are treated in a similar way in the EU program "FUTURE Baltic Sea Region Program 2014–2020 (EU Future Program 2014–2020)." Here, same as in "Post-2015 EN development agenda" is expressed strong favor in regard of shift of scientific research from interdisciplinary acknowledgment of the reality toward problematic research. There is also a social clustering for sustainable development between these researches. The methodology of social cluster analysis is common in many fields of human activity, organizing of the society and acknowledgment and management of natural processes of the nature. However, the ways of practical use are very different. Somewhat it is hard to define terms cluster and clustering. For this reason, there are so many algorithms of clustering. Smart socialization that emphasizes interaction between a person and the environment assists the process of social clustering. It is relatively divided into eight essential ways of facilitation or help (Kumpfer, Miller et al.):

1. Modeling of roles;
2. Training;
3. Advice giving;
4. Empathy and emotional reaction of caring;
5. Creation of opportunities for a meaningful participation;
6. Effective supervision and discipline;
7. Reasonable development of expectations;
8. Psychosocial or other support of other type.

Effectiveness of strategies of smart socialization, application of policy of positive actions and scenarios of positive socialization in order to establish positive components of culture and to prevent phenomena which are unacceptable by the society. Smart socialization is usually managed by professionals of social field who strive to ensure a successful scenario of every child's development and life. Namely on the basis of this theory services and assistance provided by many day centers, including children's day center "Navininkai," children's temporary shelter "Turn yourself to children," Public Institution "Education of children" positive parenthood, empowerment of a complex family, the project "Citizen." A positive result of empowerment of social communication and civil education in the process of intercultural communication can be revealed as a totality of positive actions or a positive socialization analyzed by many foreign and Lithuanian scientists (Weiner, Craighead, Guth, McGuire, Ikpa, Grusec, Hastings, Kvieskienė, Sherrod, Kumpfer and other). Namely, these three essential elements, that is, social communication, civil education and positive socialization play a special role in modeling of smart socialization. Smart socialization consists of civic education, social communication and positive socialization and act through 3D model (multifunctional, multisectorial and multi-criteria paradigm of smart socialization) which essentially unfolds in trend of smart socialization: involving and creative society.

Transparency and anti-corruption measures can be easily adapted, and goof experience can be easily recognized: as there is more transparency and open access, the more and easier we can find information we are interested in and to have equal benefit from services provided; and the public policy is more transparent. One of the best examples that can be recognized by every Lithuanian is change of the passport or Personal Identity Card or withdrawal of these documents from

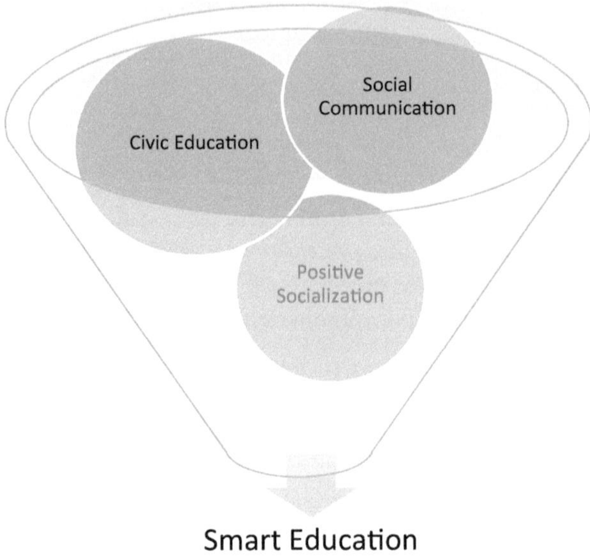

Fig. 35: Smart education

Police Department in case of their loss.[17] After investment of the European Union, this service became simple and attractive. Let us compare this service with our visit in outpatient clinic, enrollment in preschool or primary educational institution in some cities. Let us calculate resources already invested in institutions of these sectors. What else has to be done and is being done to achieve the goals posed in the project (when the starting point shows that they are the most important?). The project that is being implemented will help not only improve own competences in a more flexible way, to exchange information and own experience in easy way but also to create an operating social cluster that will assist positive social innovations.

Kvieskienė, Kvieska (2012) and Jucevičius (2009) analyze interfaces of social partnership and social clustering in other countries of the European Union as well as experiences and success cases. Social clustering, empowered regional policy would help solve social problems in Lithuania. Practical steps required for practical implementation of social partnership and social clustering in Lithuania take much time. Neither adaption of

17 Personal Identity Card. Internet Access: http://www.policija.lt/index.php?id=19572

laws neither adaptation of conception is enough for this. For example, the Law on Concessions of the Republic of Lithuania that was adapted on the September 10, 1996, No. I-1510 is applied very unwillingly. The laws on and procedure of complex family assistance same as commissions and councils of child's welfare do not promote a proper intersectional synergy either. We know how to create the description of procedure concerning educational assistance, social support and health care provided in a complex way to preschool children and their parents (custodians)[18] and many other normative documents and to this end use resources of European funds and national resources. However, little things change in a real life. While strengthening a role of own organizations every individual will be able to improve in a flexible way own competences that are necessary for a successful social clustering. It helps to apply current thinking about ecosystem services (what nature does for people) in line with the commonly accepted goal of 'sustainable development. It is the primary framework for action under the international Convention on Biological Diversity (CBD). Successful communities, schools and regions are based on smart education, innovative and multifunctional design of environment, public spaces, which include social partnership between public, private and civic organizations. The ecosystem approach is a holistic and inclusive look to the natural environment. It helps to apply current thinking about ecosystem services (what nature does for people) in line with the commonly accepted goal of 'sustainable development. It is the primary framework for action under the international Convention on Biological Diversity (CBD).

Interactive strategies of gamification, one of industries of technologies rapidly gaining popularity in the world, should be used when scenarios of positive socialization are applied. It is forecasted that until 2020 more than 40% of all science and works in the world will be associated with it. Already today a game "Pinigenai" www.pinigėnai.lt; www.misijapinigai can be used for family estimations and development of financial skills of children. Teachers can be encouraged to participate in virtual training: www.socin.eu, together with children they can develop projects and deal with social problem,[19] discuss

18 The 4th of November 2011 No. V-2068/A1-467/V-946

19 Project Citizen program overview. Internet Access: https://www.youtube.com/watch?v=QxocppEVoB8; https://www.youtube.com/watch?v=V08cKOJQk5c&list=PLE31CCC840FF35933&index=2; Stages of the Project Citizen. Access through the Internet: http://www.lduk.lt/apie-lduk/ateities-pilietis/projektas-pilietis/;

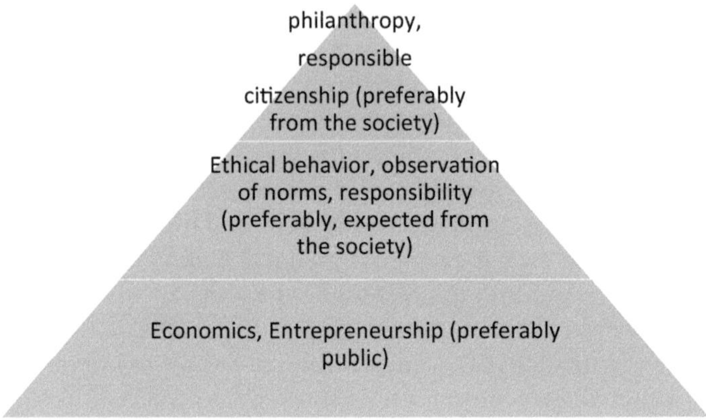

Fig. 36: Ecosystem approach. Ecosystem Approach. Internet Access: https://ecosystemsknowledge.net/ecosystem_approach

Fig. 37: A pyramid of smart education

Future citizen. Access through the Internet: https://www.youtube.com/watch?v=ur-Dz9QjVUg; Organizing of Aflatoun program implementation in regional level. Access through the Internet: http://www.kpmpc.lt/kpmpc/projekto-svietimo-darbuotoju-kvalifikacijos-kelimas-per-mokymasi-visa-gyvenima-motyvacijos-didinima-interaktyviai-kuriant-ir-diegiant-ugdymo-programas-nr-eee-lt08-smm-01-k-02-0/; www.pinigėnai.lt; www.misija.pinigai.lt

about posters or pictures created during teamwork, for example, social business was visualized in the business forum of 2017 by partially applying gamification, simulation. The forum of social business in Lithuania was first organized in 2014. The guests not only from Lithuania but also from abroad take part in the forum every year. The main tendencies of social business and its growth (both in European and Lithuanian level) are discussed here every year. The participants also discuss things that should be ensured in order to create efficient social partnership, efficient provision of public and social services.[20]

There remains a wide unbroken soil for those who are looking for acceptable financial mechanisms and mutual agreement. To trust means to allow acting during implementation of own way of thinking. Social businesses are the most organic when they arise from the need of a particular leader, deliberate choice and determination for that.[21] Social business in many countries of the world where its development was started quite long time ago is developed sufficiency well. In Lithuania, the prepared Conception of social business is seen as the first step in development of social business.[22] The Conception is designed to implement in Lithuanian the recommendations of the Communicate No. (2011) 682 of the European Commission "Initiative of social business. Creation of environment that is favorable for development of social companies in the economics of social innovations" dated on the October 25, 2011, and the Resolution of the European Parliament "Initiative of social business". Since the terms "social business" and "social companies" used in the Communicate No. (2011) 682 of

---

20 The Social Business Forum. Internet Access: https://www.google.lt/webhp?sourceid=chrome-instant&ion=1&espv=2&ie=UTF-8#q=sumaniosios+socializacijos+ir+edukacijos

21 Giedrė Guodienė. Internet Access: https://www.facebook.com/photo.php?fbid=729099557260913&set=a.216568108514063.1073741829.100004826252816&type=3&theater

22 The Minister of Economy of the Republic of Lithuania. THE ORDER REGARDING AMENDMENT OF THE ORDER NO. 4-207 OF THE MINISTER OF ECONOMY OF THE REPUBLIC OF LITHUANIA "REGARDING APPROVAL OF THE CONCEPTION OF SOCIAL BUSINESS" DATED ON THE 3RD OF APRIL 2015. Internet Access: https://ukmin.lrv.lt/uploads/ukmin/documents/files/SVV/Soc_%20v_%20koncepcijos%20pakeitimas%202016-08-29.pdf

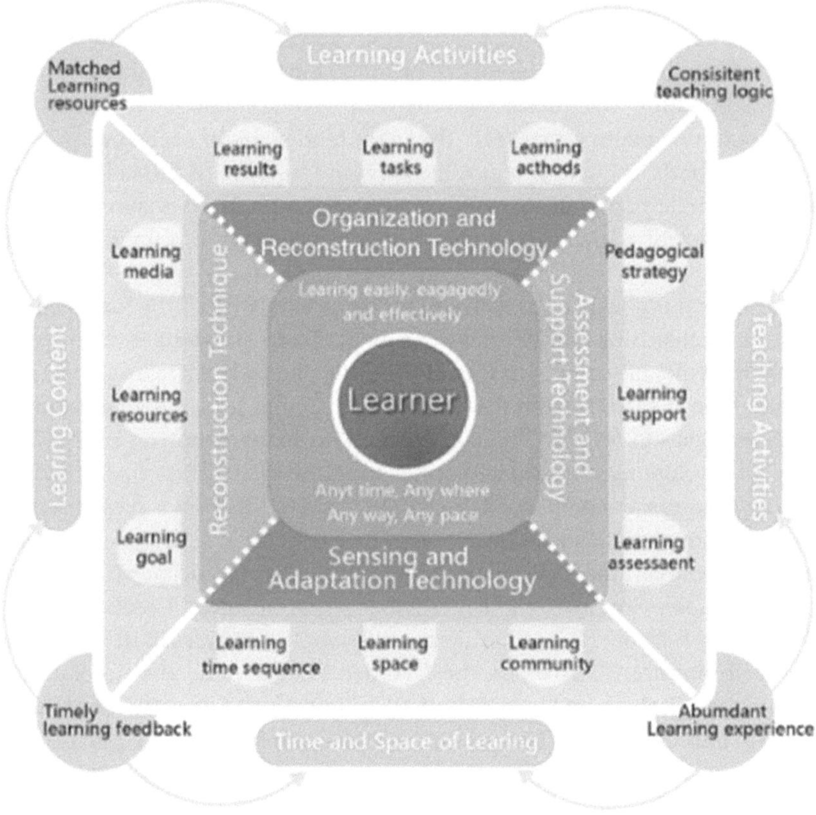

Fig. 38: Smart learning environment

the European Commission "Initiative of social business, creation of environment that is favorable for development of social companies in the economics of social innovations" dated on October 25, 2011, for this reason only a term "social business" is used in the Conception, because it is more general.[23] An

23 Social business conception. Internet Access: https://ukmin.lrv.lt/uploads/ukmin/documents/files/Verslo%20aplinka/socialinis_verslas/Socialinio_verslo_koncepcija_2015_isakymas.pdf

Social Business Initiative (SBI) follow up: Co-operation between social economy enterprises and traditional enterprises Final Report. Internet Access: https://ec.europa.eu/docsroom/documents/36684

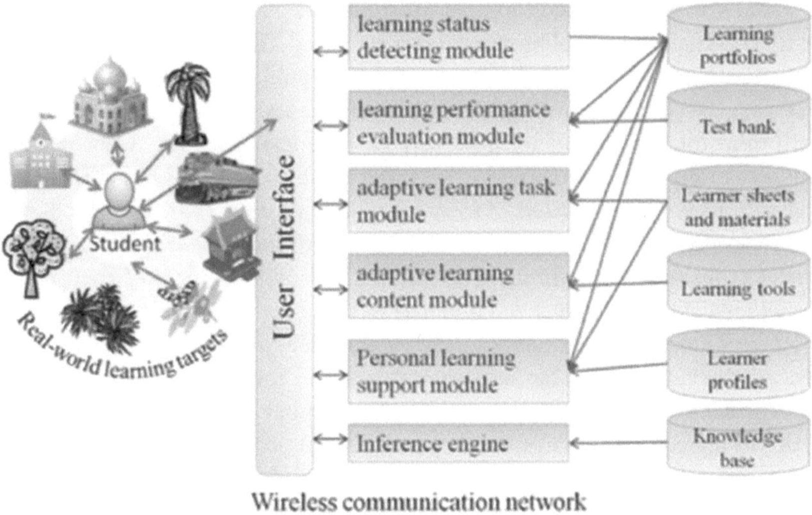

Fig. 39: The impact of ISO 9000 and ISO 14000 on standardization of social responsibility. Pavel Castka, Michaela A. Balzarova (2008).The impact of ISO 9000 and ISO 14000 on standardization of social responsibility—an inside perspective. International Journal of Production Economics 113(1):74–87. Internet Access: https://www.researchgate.net/publication/222042022_The_impact_of_ISO_9000_and_ISO_14000_on_standardisation_of_social_responsibility-an_inside_perspective

important model of the social economy is the social enterprise also known as "social business", "social-purpose business" and "social-driven business". Social enterprise is broadly defined as an organization pursuing a social mission, trading to tackle social problems and creating benefits for the society or the environment (Haugh, 2006; Defourny & Nyssens, 2008, 2016). Preparation of the Law on Social business was started; discussion regarding legal regulation and support of the state in development of social business was started. In the United Kingdom social business is granted free of charge premises of municipality.[24] The same cannot be done in Lithuania as the Law on Property disposition prohibits this. Hence, there is a need to consider corrections. This will pose a certain challenge to municipalities and some state institutions because

24 Internet access: http://www.delfi.lt/projektai/lietuva-rytoj/delfi-naujienos/verslai-kuriuos-uzsienyje-jau-gliaudo-o-lietuvai-tai-dar-kietas-riesutelis.d?id=68656422

all social services provided by them must be reviewed and new opportunities for their transfer to social business must be found. Social companies have the particular benefits of income tax or benefits of public procurement. It is also necessary to consider granting of social benefits to social business. These things are important and they determine why so far social business is not popular enough in Lithuania. We should expect that these goals would be achieved.

A summary of interactions of smart socialization and smart education gives rise to development of change of traditional way of thinking as the most important phenomenon and optimizing of public sector, development of civil society, phenomena of social partnership and networking.

Student comments are interspersed to illustrate the changes students experienced through the course of a semester social industries (social business, social innovations, social economy) in the member states of the EU operate in the multisectorial environment and the main ecosystems of social industries[25] serve as active social inclusion, services, education of children and young people and social development of remote locations. Social industries are such activities which are based on individual's abilities of smart education and social communication to gather together a team, community, social network for the general goal by changing welfare of an individual and his environment. The goal of activities of social industries is to develop services or products based on social innovations or intellect which free creative abilities of an individual, create a material welfare, contribute to creation of jobs and ensure a greater partnership and synergy of companies, organizations and sectors (social capital). "Search for competitive social market, territorial cohesion, original solutions for public problems especially in fight against poverty and exclusion is described in the strategy "Europe 2020," in the example initiative "The unit of innovations,"[26] in "The plan of Europe's fight against poverty and social exclusion and "In the Act of the general market."" Interfaces of smart education and social industries with welfare of children and the society are analyzed by G. Kvieskienė (2003, 2005, 2008), G. Kvieskienė, V. Kvieska (2012),

25 What is the ecosystem approach? Internet Access: http://www.projectpisces.eu/guide/implementing_the_ecosystem_approach/what_is_the_ecosystem_approach/

26 The example initiative of the strategy "Europe 2020" "The Union of innovations," Internet Access: http://eur-lex.europa.eu/LexUriServ/LexUriServ.do?uri=COM:2010:0546:FIN:lt:PDF

G. Kvieskienė, E. Celešienė (2014), V. Rutkauskas, (2014), D. Bardauskienė (2014) and other.

Ever changing world, change in social, economic, environmental field and values, challenges posed by terrorism, problems of refugee integration and other problems make us develop innovative, creative, ecological, and civilized society. Technologies of smart education enable social industries and social partnership, teach entrepreneurship and turn ideas into creative goods or services that offer specific benefit for communities and individuals. These strategies are taught by social pedagogics, one of the youngest branches of science not only in Lithuania but also in the world that has assumed challenge of holistic educational theory in the monograph of Professor, Habilitated Doctor Juozas Vaitkevičius[27] who has continued ideas Stasys Šalkauskis[28] (Kvieskienė, 2003) and naturally became a prototype of smart education. Bronislovas Bitinas states that social education is cultivation of a child sort of young plumed and his fostering, protection against dangers faced by him in the surrounding environment (Bitinas, 2004).[29] According to the theory of William Strauss and Neil How, a new unique generation of people that forms during a distinctive period marked by crises, cultural rise or technological break-through grows during up every 20 years. All these things determine features of human attitude and personality. Finland's education system begun to change in the early 1970s with new national framework curriculum and guidelines; teacher education was transferred from teacher colleges and seminaries to universities and comprehensive education structure was unified and schools were restructured. These factors changed the teaching and learning in Finland, subjects and skills like mathematics, science, technology, problem-solving, teamwork, creativity and interdisciplinary studies kept on greater importance in Finnish curricula (National Center on Education and the Economy). In the 1980s, National Board of Education (NBE) created new curriculum that focused on decentralized education system and this curriculum gave education providers (municipals, schools) increased freedom even the common core curriculum and guidelines for classroom hour distribution maintained still in National Board of Education. Decentralization continued through the 1990s because of

27 Vaitkevičius J. The basics of Social Pedagogy. Vilnius: Egalda, 1995.
28 Šalkauskis S. Public education. In: Selected writings. Book 2. Vilnius: publishing centre, 1992.
29 Bitinas B. Social education. In: Hodegetics. Vilnius: Kronta, 2004, p. 155–164.

Tab. 10: Western and Finnish Educations System

General Western Model	The Finnish Model
Standardization Strict standards for schools, teachers and students to guarantee the quality of outcome	**Flexibility and diversity** School-based curriculum development, steering by information and support
Emphasis on literacy and numeracy Basic skills in reading, writing, mathematics and science as prime targets of education reform.	**Emphasis on broad knowledge** Equal value to all aspects of individual growth and learning: personality, morality, creativity, knowledge and skills
Consequential accountability Evaluation by inspection	**Trust through professionalism** A culture of trust on teachers' and headmasters' professionalism in judging what is best for students and in reporting of progress

the economic recession of the early 1990s (National Center on Education and the Economy). During the last decades, the development of the Finnish education system has largely followed other Western countries but still we can find two clear differences in Finnish education system (Tab. 10):

1. There is no strong version of consequential accountability with national testing;
2. Standards are relatively open to local flexibility and diversity with a strong emphasis on basic literacy and numeracy concurrent with a wide-range education for all (Hautamäki, Karjalainen, & Kupiainen, 2008, p. 12).

The Finnish education system includes both state controlling and autonomous elements. The government and the Ministry of Education are responsible for all public education in Finland determining the general objectives of education and drafting legislation and government decisions and budgeting. The National Board of Education is responsible the core curriculum and its directive norms. Local authorities are responsible for the practical arrangement of schooling and compose the municipal curriculum. In addition, each school writes their own curriculum together with teacher staff based on national core curriculum and the municipal document (Hautamäki et al., 2008, p. 16; Finnish National Board of Education, 2015, p. 9–11).

Therefore, we have selected multifunctional, multi-criteria, multisectorial (3D/3M) smart education associated with individual need of each person and opportunities of mastering information, enabling self-help, social partnership, sharing, subsidiarity and clustering (networking) as the conception of resort of smart education. I have to acknowledge that from the first sight, the conception of smart education constructed in such way becomes complicated and hard to understand and for this reason, I get comments of my colleague's scientists and even from those practitioners whom I consider the pioneers of smart education[30]. The strategy of positive socialization constructed on the principle of 3D/3M (multifunctional, multi-criteria and multisectorial) model unites active involvement, social enterprise, positive socialization and systemic comprehensive assistance for welfare of persons, families, social groups and communities. This innovative practice already delicately sprouts not only at the schools managed by experienced leaders but also in active communities, multifunctional centers[31], innovations of social business[32]. Before modeling new services, modernizing socioeducational, multifunctional, culture, health and other public institutions, strengthening partnership of public and private sectors we have to learn reorganize spaces that are not used, enable smart education that would promote emergence of new products of creative and social industries, would initiate new services, would optimally use an infrastructure, ecological natural environment, public spaces, would integrate local traditions and global innovations through partnership of socially responsible institutions, by increasing interinstitutional social trust and developing a social capital.

One of possible principles of modeling in order to ensure a constant interaction and synergy of different sectors between public, private, NGO, science sectors and marketization of services and new products is 3 D model of smart education (multifunctional, multisectorial, multi-criteria). We find success cases of smart education in Lithuania, Belgium, Holland, Denmark, Canada, the United Kingdom, the United States of America, Ireland and other countries. Analysis of success cases confirms that multifunctional, multi-criteria

---

30 Danutė Puchovičienė, the principle of Vytautas Magnus Gymnasium, Internet Access: https://www.facebook.com/VilniausVDG/; http://kultura.lrytas.lt/istorija/lietuvoje-seniausia-gimnazija-glumino-drausme-ir-grieztumu.htm; Miša Jakobas, the principle of Jewish gymnasium, Internet Access: http://www.15min.lt/naujiena/aktualu/lietuva/vilniaus-s-aleichemo-gimnazijos-direktorius-misa-jakobas-ar-as-kaltas-kad-baigesi-vasara-56-525210

31 Sudeikiai multifunctional centre, Internet Access: https://www.facebook.com/sudeikiumokykladaugiafunkciscentras/?fref=ts

32 Social business and social innovations. Internet Access: https://www.facebook.com/groups/193744704006846/?fref=ts

and multisectorial design of institutions and organizations helps not only neutralize segregation and stigmatization of various social groups (ethnic, religious, and social, based on age) but also increases involvement, encourages creativity and innovations. Multifunctional design, social communication and partnership facilitate marketing of services and goods and increase growth of intersectional social capital. Smart education is the most receptive in terms of creative and social industries because it can be most easily realized through a positive socialization, experiential education, scenarios of experience, adventures, dreams, playful and smart education that are characterized by four main features:

- Commitment – to assume responsibility and to be open to changes.
- Objection – to focus on tensions between the actual and desired practice.
- Conflicts – to use energy for positive solutions.
- Social Partnership and clustering.

Social Pedagogy has naturally become the basis of conception of smart education because it is one of the youngest branches of science not only in Lithuania but also in the world. Therefore, it has naturally assumed challenge of holistic education whereby we also substantiate the conception of smart education. The pioneer of Social Pedagogy Professor Doctor Juozas Vaitkevičius[33] who continued ideas of Professor Doctor Stasys Šalkauskis[34],[35] and Professor Doctor Bronislovas Bitinas in his monographs by establishing their ideas and stating that social education is cultivation of a child sort of young plumes and his fostering, protection against dangers faced by him in the surrounding environment. Bitinas (2004)[36] has established the basics of holistic and at the same time individualized theory. We find success cases of smart education in Lithuanian, Dutch, Danish, Norwegian, Canadian, Irish, American, Singapore, Finish and other countries socioeducational institutions and non-governmental organizations. Success case only confirm our assumption that multifunctional, multi-criteria and multisectorial (3D) socioeducational design of organizations helps create a new quality of education that we identify as smart education. Multifunctional hubs, open centers of education and youth and spaces

33 Vaitkevičius J. The basics of Social Pedagogy. Vilnius: Egalda, 1995

34 Šalkauskis S. Public education. In: Selected writings. Book 2. Vilnius: publishing centre, 1992.

35 Kvieskienė G. Socialization and child welfare. Vilnius: Vilnius Pedagogical University, 2003. Internet Access: biblioteka.vpu.lt/elvpu/37059.pdf

36 Bitinas B. Social education. In: Hodegetics. Vilnius: Kronta, 2004, p. 155–164.

of family gamification facilitate creation and development of services and products, increase trust and synergy of different sectors and promote growth of social capital.

Demand for smart education in different sectors is determined by various tendencies: demand for active and creative recreation increases, educational system becomes more varied, net income grows, and on the other side alienation of family member's increases, parents have less hours for leisure, factors of virtual environment and other factors encourage us reconsider traditional educational methods, to reassess a quality of services and traditional education. For formation of its socioeducational environment Lithuania has chosen a modern strategy of welfare states – it strives to develop a stable social security system (social welfare) and to provide professional community oriented socioeducational services. Demand for multifunctional specialists (social educators can be identified as such) grows not only in educational institutions but also in communities, non-governmental organizations, business companies. Social educator is a professional of new generation able not only help children better adapt in the society, community, education or care institution, other social institutions performing formal and informal educational functions but also able to perform functions of career counselor and mediator, to help parents perform their duties responsibly and creatively. The principle of participating research is observed at the time of preparation of young professionals. We understand that Z generation appreciates variety in every sense and that it is technologized generation able to find easily the most complex information. For this reason, we distinguish the following priorities of smart education: flexible, database learning and work schedule, social and inner communication, personal freedom. Therefore, while applying other technologies of social innovations, such as social partnership, positive socialization, community oriented services, we have developed the innovative program of post-graduate studies of Social industries and communication in which together with post-graduate alumni engaged in practical activities we try to search for new strategies which could democratize the procedures of knowledge generation and could initiate systematic socioeducational changes. At this point we are assisted by Lithuanian Association of Social Educators[37] that establishes identity of social educators and is sort of umbrella protecting against "cold hail" posing a heavy bargain for "fresh" social educators. In their many workplaces, social educators have become prophets of social business and creative as well as social industries and

37 See: Access through the Internet: www.lspa.lt

Fig. 40: A diamond of social pedagogy/smart education (according to ThemPra's Diamond Model (Eichsteller & Holthoff, 2012)

initiators of social innovations. Therefore, we hope that new changes of conception will help in development of socioeconomics, social industries and creative innovations in communities, socioeducational institutions and NGOs.

ThemPra's Diamond Model (Eichsteller & Holthoff, 2012) symbolizes one of the most fundamental underpinning principles of Social Pedagogy: there is a diamond within all of us. As human beings, we are all precious and have a rich variety of knowledge, skills and abilities. Not all diamonds are polished and sparkly, but all have the potential to be. Similarly, every person has the potential to shine out – and Social Pedagogy is about supporting him or her in this. Therefore, Social Pedagogy has four core aims that are closely linked: well-being and happiness, holistic learning, relationship and empowerment. These goals can be achieved using technologies of smart socialization and education and a platform of social clustering. We can expect sustainable development. Smart education is sort of tools of creative and social industries that help every person to improve, to enable himself, to help others. They also help different social groups to polish all edges of a diamond of their personality in practice through 3D: multifunctional, multi-criteria and multisectorial model. In the theoretical discourse since 2009 both practitioners and scientists from Lithuania and the rest of the world, discuss by conceptualizing problematics of social clustering, social business, entrepreneurship, and social communication. More than 450 international partners from all over the world participated with their presentations in the international symposiums "Social innovations and partnership for development of social cohesion," "The theory and practice of social education (learning): tendencies, challenges and opportunities," "Positive socialization, enterprise and social responsibility" and other that were held during the period from 2009 to 2014. In majority of the countries of the EU social entrepreneurs, professionals of social industries and communication, social educators, youth, street teachers, social mediators are prepared in multisectorial environment. Today many public institutions demand professionals of holistic profile able to comprehend a multifunctional design. Social educators perfectly adapt their competencies in institutions of social education and sociocultural institutions, institutions of culture and tourism, multifunctional centers, creative factories, NGOs, Hubs, at schools of smart socialization, talents, success, democracy, survival, adventures, family recreation and entertainment camps, entertainment centers, open spaces of multifunctional design, workshops of reprieve, creation, consulting, animation and art, etc.

More and more nontraditional socio-creative services and businesses that operate under multifunctional principle in multisectorial environment appear in our country, that is, multifunctional centers, culture factories, arts printing houses, beehives of NGOs, Hubs, centers/homes for elderly people, social enterprises that integrate socially vulnerable persons, etc. These phenomena of social industries and creative innovations erase boundaries that long time divided creation, art, sector of community services and educational sector, industry, social innovations, recognize worth of a brand, create value added for identity of the country and the city, strengthen city and rural communities. We identify this as tools of smart education, which must be handed over to students in order they could learn to present themselves and their community, in order they could increase attractiveness of our country and a quality of life.

Demand for creative and social industries, services of social communication in different sectors are determined by various tendencies: demand for active and creative recreation grows, educational system becomes more varied, and Z generation poses new challenges for parents and educators. A growing net income, alienation of family members, a decreasing number of odd moments and other factors promote distribution of value priorities of the society, influence increase of demand for goods and services related to social industries, social communication, culture, social innovations and services all around the world. Change of new habits and tendencies of consumer culture promotes changes of entertainment and service industry, influences occurrence of new multifunctional tendencies in cultural and daily life, development of green ideology, creative and green communities, change of fine and performing arts, education and a daily lifestyle, individualization of new products and services and growth of value added created by the said fields, emergence of new jobs and ever stronger contribution to the gross domestic product. For formation of its socioeducational environment Lithuania has invoked the principles of decentralization of services and formation of a stable complex systems of social assistance by concentrating resources and efforts of social, educational, health sectors, and civil society. New youngest professionals – social educators – are able to coordinate a process of a complex assistance and management of a specific case. People start to realize that this "craft" is especially important both for the country's citizens and the state itself. Social educators are prepared in Lithuanian University of Educational Sciences in accordance with participation action research,[38] which was established in the studies program of social educators as technology of social innovations. Such innovative technologies as social partnership[39], positive socialization, community oriented sustainable development, learning by doing are characterized by active social inclusion. They democratize the procedures of generation of knowing and can initiate systematic social changes. The researchers Sigita Burvytė,[40] Angelė Kaušylienė, Giedrė Kvieskienė, Loreta Žadeikaitė,[41] Irena Zaleskienė,[42] Kęstutis Ralys[43] and young scientists Eglė Celešienė,[44] Rokas

---

38 Jarg Bergold & Stefan Thomas. Participatory Research Methods: A Methodological Approach in Motion. FQS.Forum:Qualitative Social Research.2012, No. 1, Art. 30 – January 2012, Access through the Internet: http://www.qualitative-research.net/index.php/fqs/article/view/1801/3334

39 Kvieskienė, G., Kvieska, V., (2012) Influence of social partnership on innovations. Monograph: Educology. Vilnius, 265 p.

40 The greatest miracle - life. In memory of Professor Zenonas Bajoriūnas, composed by S. Burvytė. Vilnius: Lithuanian University of Educational Sciences. 2012

41 Žadeikaitė,L., Gulbinas R., Improvement of socio-educational environment of social educator using ICT. Access through the Internet: http://socialinisugdymas.leu.lt/index.php/socialinisugdymas/article/view/38/34

42 Zaleskienė, Irena. Youth citizenship: between text and context/Nationality and citizenship. Exclusion or harmony?: monograph. Vilnius: Publishing house of Vilnius Pedagogical University, 2007.

43 Ralys, K., Preparation for marriage and family: scientific monograph; Vilnius: Publishing house of Vilnius Pedagogical University, 2011.

44 Kvieskaitė/Celešienė E., Expression of interfaces of intercultural communication and educational innovations: doctoral thesis: social sciences, educology (07 S)/Eglė Kvieskaitė; Vilnius Pedagogical University. – Vilnius: Publishing house of Vilnius Pedagogical University, 2011. – 133 p.: iliustr. – The thesis has been prepared in

Gulbinas,[45] Renata Katinaitė and Giedrė Misiūnienė[46] establish traditions of their teachers Juozas Vaitkevičius, Bronislovas Bitinas, Zenonas Bajoriūnas.

A persistent number of children living in care institutions, with one parent or grandparents, children whose one parent lives in a separate household or abroad, bulling and violence, extent of suicides, a number of support professionals currently available in educational institutions is at least three times smaller than a number required to work effectively. Therefore, problems of socioeducation and positive socialization in the country still are relevant. In the modern society, there is a lack of systematic (holistic) attitude in all fields of a public life. According to Professor Dr. Meilė Lukšienė, the pioneer of Lithuanian reform of education, the modern society should establish this attitude in education. The Professor states that education is much more than things happening at school, that education is a holistic process which realizes a potential of person 's learning and growth which can happen in any situation and when a social educator moderates in educational environments and is a lifelong process that involves "the head, the heart and the hands." However, a creative teacher is a leader of this process.[47] Next to holistic attitude, when smart specialization is emphasized in Europe, we want to substantiate that smart education, that initiates and expands the process of cognition, helps young people to free their creative potential and that is effective in social-ecological community by involving all sectors and activities available in the community in which social industries can form or can be initiated, is not less important.

Vilnius Pedagogical University in 2006–2011. – Doctoral thesis (social sciences) – Vilnius Pedagogical University, 2011.

45 Gulbinas,R., Pedagogical substantiation of educational environment of social educator enriched with information and communication technologies. Doctoral thesis: social sciences, educology (07 S)/Rokas Gulbinas; Lithuanian University of Educational Sciences. – Vilnius: Educology, 2014. – 229 p.: iliustr. – The thesis has been prepared in Lithuanian University of Educational Sciences in 2010–2014. - Doctoral thesis (social sciences) – Lithuanian University of Educational Sciences, 2014.

46 Misiūnienė, G., Modelling of socioeducational activity of multifunctional centre in the rural (territorial) community): doctoral thesis: social sciences, educology (07 S)/Giedrė Misiūnienė. Lithuanian University of Educational Sciences. – Vilnius: Educology, 2014. – 229 p.: iliustr. – The thesis has been prepared in Lithuanian University in 2010–2014. – Doctoral thesis (social sciences) - Lithuanian University of Educational Sciences, 2014.

47 Positive socialization, entrepreneurship and social responsibility. Access through the Internet: http://www3.lrs.lt/docs2/DHXNYELG.PDF

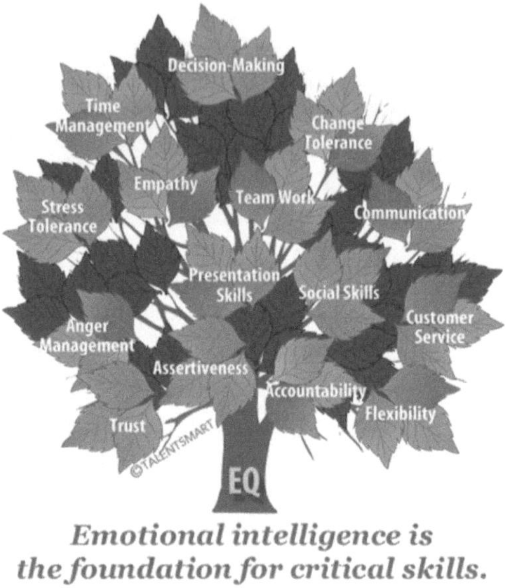

Fig. 41: Cultivation of social partnership throw trust

Smart education is a tool of creative and social industries that helps every person to improve, to enable himself and to help enable different social groups. This trend of smart education operates under the principle of 3D/3M – multifunctional, multi-criteria and multisectorial – model, multisectorial scenarios of positive socialization that connect the methods of active involvement, social enterprise, smart socialization and systematic complex socioeducational assistance for those who help solve individual innovations, innovations of increase of welfare of social groups and communities and help initiate social innovations. Smart education approaches the issue from a universal design perspective. Working together with educators, parents, students, community and civic society leaders, representatives from research institution to delve into the values and systems of the social industrial model, to develop a strong sense of cognitive dissonance, and to pose highly-effective new ways of schooling that spur high engagement and high achievement for far greater numbers of students. Empowerment and re-energized the different stakeholders from different sectors to create and sustain the rich and rewarding learning environment and share it with students, the community and others civic society representatives:

- Smart Education successes employ principles of universal design innovation that help us work with you to re-imagine and then build the kind of organization and culture you need to move forward. Key school re-design principles we espouse and use with you to custom-build your new school community include, empowerment all community sectors;
- Maximizing community connections and partnerships that build on permeability to ideas, resources and opportunities;
- Creating a culture of learning, through gamification and project making rather than a culture of teaching, built upon research-based notions of authentic, intellectually challenging work;
- Building strong social-emotional supports and resilience for all learners, their families and educators;
- Re-imagining job descriptions and functions to support new social education innovation prototypes;
- Using the right data at the right time, including qualitative approaches that inform and drive improvement rather than measurement and compliance;
- Employing the new wealth of research from learning neuroscience, motivation and child and adolescent development and organizational analysis.

In most countries of the EU social entrepreneurs, professionals of social communication, social educators, youth, street staff and professionals of social and creative industries are prepared to work with children, youth, adults and different social groups and families. They adapt competences in the sociocultural institutions, multifunctional centers, creative factories, NGOs, Hubs and art workshops and in different services of public, private and NGO sectors, in business where provide services of entrepreneur, coacher, youth consultant, street teacher, coordinator of social assistance, social mediator, animator, facilitator under the principles of multifunctional, multi-criteria and multisectorial design. In order to prepare various social grows op for partnership, these professionals usually apply interactive gamification strategies. The origins of game literacy can be found in the tradition of information literacy that was later expanded to include such areas as media, digital literacy and social communication (Buckingham, 2007). Digital literacy becomes more and more important because the society more and more often visits and obtains information placed in the digital medium. Digital literacy is important but not the only competence that plays a role when striving for development of gamification scenarios. According to the data of Entertainment, software association (2014) casual games represent more than one third of all games played. Educational scenario is rarely relied on in entertainment production of this type. However, elements

Fig. 42: Assumptions of game literacy. Adapted according to: Game literacy at GameIT College. Poulsen, M., Introduction to game based learning//THE GAMEiT. Edited by: Mathias Poulsen and Ebba Køber A framework of game based learning pedagogy. 2011. Access through the Internet: http://www.mathiaspoulsen.com/gameithandbook.pdf

of game design that promote involvement in activity and that are used both in gamification educational platforms and in entertainment, games are the objects of education rather than the objects of entertainment. We can distinguish eight assumptions of game literacy in order of priority. They were distinguished by seven national and international experts surveyed by us:

- Creativity and innovations (1);
- Entrepreneurship (2);
- Development of virtual games (3);
- Social communication (4);
- Cooperation, partnership (5);
- Analysis, reflection (6);
- Information literacy (7);
- Determination, opportunities (8).

Game design and construction of scenario takes place by applying the principles of design and aesthetics in order to create interaction between education of gamers, business simulation, entertainment or medicine, or for experimental purposes. Game design can be used both when creating games for entertainment and when seeking notional gamification interactions. Gamification process requires posing goals, rules, to provide challenges, to define a type of game: simulation games, games of formation of practical, social skills, video, roles, casino, modeling games creating desired interactions between the participants, or maybe between spectators. In terms of academic analysis, design of a game is a part of game theory that analyzes strategic decisions (first of all in situations other than situations of gamification). Elements of design or scenario can be various: detailed descriptions of all elements or a detailed program, systems of levels and means of reward or tables of leaders. In case of gamification, elements of design, dynamics and aesthetically rendering of content are only means that help to consider the most important tasks by promoting interest in activity and involvement in the process of tasks solving.

Games often provide an opportunity to try various scenarios of acting within frames of rules and restrictions. Every game is based on the set of particular rules that has a limited number of restrictions and templates of dynamics. It is much easier to meet its requirements than to solve challenges in a real life. Gamification can come in handy as a method of "supplement" of the reality.

Curiosity arouses and gamers are stimulated to seek for a new quality of acting when a layer of game is adapted for performance of daily tasks. So, how we could define the main principles of gamification, which could help, expand spaces of learning and define an essence of gamification? Based on the authors discussed above and Mathias Poulsen (2011)[48], we distinguish five main principles of gamification:

- Problem solving (1);
- Context of acting (2);
- Feedback (3);
- Optimal challenges (4);
- Comfort zone (5).

48 Poulsen, M., Introduction to game based learning//THE GAMEiT.Edited by: Mathias Poulsen and Ebba Køber A framework of game based learning pedagogy. 2011. Access through the Internet: http://www.mathiaspoulsen.com/gameithandbook.pdf

A game is the simplest tool those teachers to solve problems (Gee, 2007, Corbett, 2010). The authors who point this out note that problem solving is a core of all games (1). A character of these problems is various, as creative games of early childhood colored with colorful fantasy, not less colorful than search for "Pokémon's" that overwhelmed minds of young people all around the world. Rich reflections of a virtual world in video games have powerful action contexts (2) which can be useful for overcoming complex tasks of learning. Worlds of games help connect words and symbols, understand better meanings and specific situations, and experience real situations. Through properly selected educational games experience of earlier activity is reproduced in several contexts, the learning individuals understand much easier complicated terms without losing abstract interactions of ideas with solving of real problems. Thus, games, especially those that simulate a virtual reality, are strong educational means allowing improvement and expansion of personal challenges and opportunities of perception (Shaffer, Squire, Halverson, & Gee, 2005). Learning is often hard because we do not see feedback that comes only after some time expressed in the form of grade and, at best, in the form of niggardly comments. Whereas after gamification interim evaluations that are happening faster in the way of encouragement and support inevitably become more, frequent.

When we solve problems in the game and constantly get a positive feedback, the latter also specifically measures and evaluates our contribution. Whether we search for "Pokémon's," fight by chasing enemies in the outer space, create cities or draw up family budget we are always praised and gain growing points of evaluation or prizes. It doesn't matter whether we get response in a form of a message, a dialogue, points, in a form of meters we have walked, sounds of music or vibration – all these irritants have a positive effect. Their effect differs from an effect of school bell that invites us to a lesson we have not prepared for.

Gamification strategies are resources that have not been yet mastered in educational practice, which can help educators to seek their own positive socialization and sustainable social partnership and might help individuals seeking for new knowledge to achieve optimal results playfully and with positive attitude. Gamification strategies are not panacea. However, they provide both children and adults with a chance of overcoming stress and tension when striving to gain additional skills and competences. When creating gamification scenarios and strategies we should not forget monitoring of this process and evaluation how one or another gamification scenario has involved players and has achieved the goals posed by us. Positive socialization – purposeful actions of competent persons in order to achieve welfare of a child and his next of kin. However, when seeking for this welfare, it is not enough to have a coordinated effect of a family,

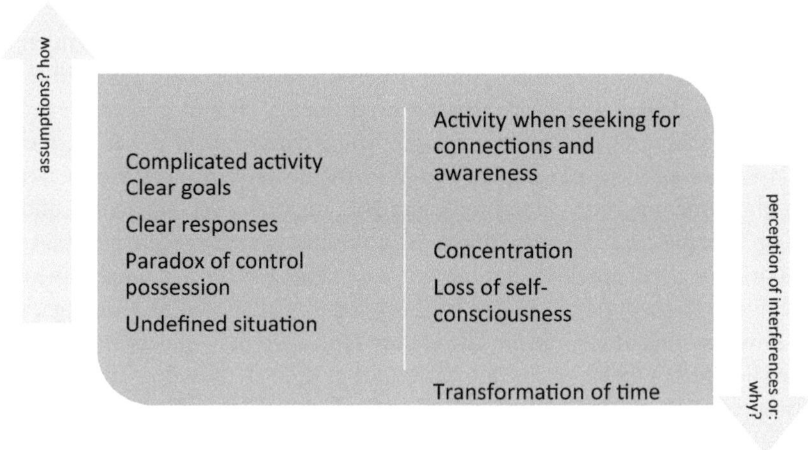

Fig. 43: Assumptions and interferences of successful gamification

school, community, NGO. Smart education poses four main goals, which are associated with personal welfare and happiness, holistic education, partnership and empowerment, and seeks that all edges of this diamond would be equally polished in practice through 3D model: multifunctional, multi-criteria and multisectorial effect when cultivating a social clustering, when seeking for sustainable social welfare and enables individuals, community and civil society. One of the tasks of smart education is to develop skills of social communication and social clustering of a full-pledged person.

Education theories constantly analyze goals, relations, motivation of a learning individual and learning (Vygotsky, 1978, p. 86). Mihaly Csikszentmihalyi (Csikszentmihalyi, 1990) describes the principles in the theory of "flow." Csikszentmihalyi assumes that persons who pose optimal challenges for themselves understand that these challenges are directly closely associated with existing skills and competences of a person. The theory of flow is usually considered one of the theories that mostly meet the condition of learning through games. Mihaly Csikszentmihalyi stated that caught by inspiration artists forget hunger and thirst. When they finish their work, they immediately loose interest in a fresh drawing. The same can be said about writers working on a new book, scientists who are approaching important findings, blacksmith working on a work that is important for him. M. Csikszentmihalyi defines a flow as a condition of mind when a person completely becomes involved in things done or created. In order this would happen first of all, as stated by the author, there must be a task that we perform, that is sort of challenge for us that must be solved. Theorists of gamification and designers of games have adapted these optimal challenges, which are usually used during human education in the form of "learning by doing," by posing optimal challenges (Lave & Wengeras, 1991) for we learn to participate in life of the society through gamification. This phenomenon is discussed by Johan Huizinga who established a conception of "a magic circle," for Huizinga, magic circle is a space created for playing, a material or conceptual temporary world dedicated to the act apart. However, though the concept of the magic circle is usually attributed to Huizinga, and the above passage does capture the idea behind how the concept is used even today, he did not dwell on the subject for long. Indeed, the term "magic circle" only appears six times in (the English translation of) Homo Ludens, and only three times in the chapter game scholars usually refer to: in the passage quoted above, once in the metaphoric meaning it has since acquired in game studies (Ibid, 11) and once as an example of a sacred space (as opposed to playful space, though Huizinga stresses that formally such a distinction cannot be made, Ibid, 20). However, in the context of the whole book it is not just an example or a metaphor among many, but a core feature of the examples given (as pointed out by Calleja 2012, see Huizinga 1938, 77, 210, 212).[49],[50] Homo ludens (Huizinga, 1938). According to him, a game is sort of the main feature of a human. True becomes evident after having realized that a game is an excellent mean of knowledge transfer. Moreover, although from the first sight the process seems flimsy, knowledge that is being transferred is important for further improvement of a person, creating a zone of comfort in which, like in the children's playground, we feel safe and can boldly explore the world and overcome the most complex problems without losing anything. A content can be inserted into game. Educators and parents must knowingly select and integrate into a game a required content of training. It means that in order to successfully use gamification in educational activity we have to learn how to select goals and to choose a content (Egenfeldt-Nielsenas S., 2011). When this

49 Huizinga's critics have mounted a convincing case that his conceptions of play and seriousness as well as play and everyday life are muddled: Ehrmann (1968, 32–33) questions the lack of definition for "reality" or "everyday life," and how it is possible that reality exists prior to its component, play.

50 Jaakko Stenros (2014). In Defense of a Magic Circle: The Social, Mental and Cultural Boundaries of Play. Access through the Internet: http://todigra.org/index.php/todigra/article/view/10/26

integration functions well, a strategy of gamification creates a perfect context for exploration, cooperation, consolidation of partnership and cultivation of sustainable social partnership.

A model of smart education based on 3D/3M: multifunctional, multi-criteria, multisectorial model, priority and leadership synergy of smart education that act locally and globally. Trends of break-through and new opportunities are implemented by empowering management based on data, innovations, networking and social clustering for a bigger welfare of a person and the society (community). Rationally using resources and analysis of sustainability probabilities is based on systematic complex assistance and intersectional social partnership of professionals as well as leadership of social educators, coordinating a process of a complex assistance and management of specific case.[51]

International Day of Happiness was celebrated every year on March 20th. On June 28, 2012, General Assembly of the United Nations announced the resolution whereby all member states and all organizations are encouraged to celebrate International Day of Happiness on March 20. It is noted in the resolution that pursuance for happiness is the main purpose of human. Happiness index of the countries is calculated since 2012. In order to rank countries by happiness index economic researchers usually use six factors: GDP per capita, long life duration, social support from a side of the family of friends, trust in the government, a feeling of a general personal freedom and generosity measured taking into account donation to charity organizations. It is noted that a feeling of happiness of the residents is usually ruined by inequality, poverty and corruption. Therefore, such countries used to be ranked the lowest in terms of happiness index and used to appear on the bottom of the list. Norway is the happiest state of the world. For the first time in history in 2017, it took over this title from Denmark when the United Nations (UN) announced such list. In this year Lithuania was ranked 52nd (earlier, it was ranked the 60th). Our country has outran Latvians and Estonians.[52]

Norway is a safe country because the locals feel safe enough – compared to other countries theft rate in Norway is relatively lower, the locals still leave their home doors unlocked. It is safe enough to live in Norway and a safe

---

51 Torhild Slåtto. What have we achieved in the project - some evaluation from 4 April in FuN's office in Oslo. Access through the Internet: https://padlet.com/tsb9kl/n6m9cayurc6w

52 The happiest people live in Norway. Access through the Internet: http://tiesa.com/naujienos-ir-zinios/pasaulis/laimingiausi-zmones-gyvena-norvegijoje-26709

Picture 6: Leisure time

neighborhood is strongly appreciated here. The locals do not feel stress when they stroll in the dark compared to opposite situation on the other side of the Atlantic Ocean. Public transport is well developed and safe in the largest cities in Norway. Therefore, there is no strong need to travel by car and to maintain a great physical state. Moreover, police in the streets of Norway patrols unarmed. On the other hand, some Norwegians have personal guns – even 25% of households have a gun – under the law these home held guns must be kept unloaded and cartridges must be kept under lock.

Importance of the Students Leisure Time

The students in Norway rarely complain about excessive workload and endless homework's. Moderate homework's in the country encourage development of discipline, curiosity and efficiency of learning. However, this in no way causes stress and does not deprive a no structured leisure time or a desire to participate in extracurricular activities or a meaningful community activity. Political involvement in Norway is the biggest in the world. Norwegians actively participate in activity of political parties and gladly volunteer. A difference between a full-pledged democracy and a defective democracy: a full-pledged democracy is "based on political culture that contributes to prosperity of democracy."[53]

About 78% of the citizens in Norway expressed their will during votes. It is 20% more compared to the USA. Registration as an obstacle for voting is one of the reasons of such big difference. Registration for the citizens of Norway is an automated process. Social capital reflects trust. Therefore, despite a fact that there still are various terms describing social capital, it mostly influences feeling and happiness of people.

Social capital is increase of trust of social relations (Fukuyama, 1995, Kvieskienė, 2005), civil participation (involvement) developed by participating in voluntary organizations (Putnam 1993, 2000), a structure of the society by creating willingness to participate in development of physical capital (Kvieskienė, Kvieska, 2012), an aspect of a social structure that promotes specific forms of actions and cooperation (Coleman, 1988), a rate embodies in the social relations that promotes cooperation between individuals (Fukuyama, 2001). Based on scientific research the researchers of the Legatum institute, when assessing social capital, distinguish three classical groups of conceptions of social capital, which are associated with a social cohesion, involvement (a formal volunteering, assistance to strangers, donation and trust), and civic societies (networks of communities, non-governmental organizations, families). "The Social Capital sub-index measures the strength of personal relationships, social network support, social norms, and civic participation in a country."[54] According to assessment of social capital, it consists of social cohesion and involvement (a formal volunteering, assistance to strangers, donation, trust and networks of communities and family, etc.). Lithuania is ranked 96th (2018).

Best for Social Capital

1. New Zealand
2. Australia
3. Canada

---

53 What the USA could learn from the globally best Norwegian democracy. Access through the Internet: http://skandinavija.today/naujienos/pasaulio-naujienos/ko-jav-gali-pasimokyti-norvegijos-geriausios-pasaulio-demokratijos

54 Revealed: The world's happiest, safest and richest countries. Access through the Internet: http://www.telegraph.co.uk/travel/news/prosperity-index-2016-happiest-countries

4. United States
5. Iceland
6. Norway
7. Denmark
8. Malta
9. Germany
10. Ireland

A rapid growth of economy resulted in increase in differentiation of the residents' income because a speed of growth of income of the groups of the most socially vulnerable residents was slower compared to other groups of the residents. Social partnership and innovations, a broader social partnership inspire creativity, promote a positive personal and social, ecological involvement and optimize human resources and increase mutual trust and social capital. Socioeconomics in Europe is being developed rapidly. For this reason, it is very important that social clustering, partnership, networking would promote more rapid social innovations in Lithuania.

In summary, it can be stated that influence of a social partnership on development of novelties, social innovations and investment plays a pivotal role. Partnership of several sectors always initiates occurrence of new ideas, jobs, improvement of social-ecological environment and quality of life as well as a better and more efficient way of social solutions. Based on social partnership social networks, other umbrella and network organizations increase intersectional trust form.

3.2 Success Stories for Smart Education leadership.

International Day of Happiness was celebrated every year on March 20. On June 28, 2012, General Assembly of the United Nations announced the resolutions whereby all member states and all organizations are encouraged to celebrate International Day of Happiness on March 20. It is noted in the resolution that pursuance for happiness is the main purpose of human. Pursuance for happiness and welfare is one of the purposes of this virtual learning too. Therefore, according to Morten Flaten Paulsen, the member of the Board of Virtual learning in the web the elements of encouraging of the students and elements of celebration should not be forgotten during training. FuN (acronym for a web organization "Fleksibel utdanning Norge" translated as "Flexible Education Norway") represents a web association of flexible education.[1]

SV Bybi established in **Denmark** in 2011 provides home to the bees on urban roofs and trains homeless people and asylum seekers to take care for them and to harvest honey. Thus, a social mission of this business is to provide job opportunities for long-term unemployed people and provide urban residents with an opportunity to get in contact with the nature. Bybi income consists of business agreements (25%), sales (60%) and educational, accommodation projects (15%). Businesses lease out bees and get CRS package, events and honey. Bybi does not receive any grants and 100% of its income is dedicated for implementation of social and environmental projects. Bybi has full-time employees and assistants, that is, long-term unemployed people and volunteers. This organization operates in 20 locations of Copenhagen. Paid tours are organized to Bybi.[2]

"Orange Fiber" business in *Italy* strives to solve a problem of waste generated in Italy by citrus industry, that is, 700 000 tons per year. This waste pollutes the environment and is uneconomical as costs of its management are high. Thus, two girls Adriana and Enrica created a sustainable, biodegradable material made of orange peel. The beginning of the project was the final fashion design

---

1 FUN. Access through the Internet: http://fleksibelutdanning.no/om-fun/english/

2 Neringa Stroputė, Modesta Kairytė. How social business becomes social business? Analysis of good foreign practice. Access through the Internet: http://kurklt.lt/wp-content/uploads/2015/11/U%C5%BEsienio-%C5%A1ali%C5%B3-SV-vystymosi-ir-pl%C4%97tros-gerosios-praktikos-analiz%C4%97_Neringa-ir-Modesta_2016_Kurk-Lietuvai_.pdf

Picture 7: In the photo above: A tutor of Norwegian Grant project Trost, Jana (Brussels), the president of FuN Torhild Slaato, the member of the Seimas of the Republic of Lithuania Virginijus Sinkevičius

work of Adriana Santanocito in the University of Milan (AFOL) where she analyzed the subject of new technologies in the fashion sector and production of textile. Adriana has established Orange Fiber together with Enrica Arena who completed communication studies. They shared not only accommodation but also Ecology philosophy which formed a basis for implementation of the project.

Several social companies based on the principles of social business and social clustering also operate in **Lithuania**. One of examples[3]: a company which manufactures buckwheat hull cushions operates in Vilnius. Most of the employees of the company are disabled people. The company operates in accordance with democratic principles. Unlike in the major corporations where a meeting of few people decides a fate regarding thousands of people,

3 Kvieskienė, G., New phenomenon: sprouts of socioeconimics between private and public sector. Future business. (2013, Nr.1). Science popularization, educational journal. Social liability: challenges and opportunities. Access through the Internet: http://www.svak.lt/images/ateities_verslas-spauda_ok_web.pdf

its members have the right to express their will during the process of decision-making. Another inspiring example: a coffee shop "Mano Guru" operates already eight years in the capital. Addicted people undergoing treatment work in a stylish restaurant-salad bar operating in the central street of the city. Nevertheless, it is obvious that the establishment has difficulties of competing with other restaurants established in the same street. However, with a little aid of the state it can survive and create welfare for the society – after having received a little bit of financial injections from the state the coffee shop integrates vulnerable people to the society, teaches them working skills. Only business or only the state could not achieve such results. Skeptics may say that non-governmental organizations already operate in Lithuania because public groups representing individual sectors can be found in every municipality. But practically such groups have no real impact on the Board. Such organizations can be identified as sham accompaniments to the city Boards. If public organizations and subjects of socioeconomics were more influential, then more social and economic problems would have been solved already. The authorities should be more determined and rely much more on Lithuanian people. They should only be provided with appropriate tools. Authorities should check how money is being spend and support people in case of extended funding if people execute their plans and effectively use support provided to them.

International Day of Happiness was celebrated every year on March 20. On June 28, 2012, General Assembly of the United Nations announced the resolution whereby all member states and all organizations are encouraged to celebrate International Day of Happiness on March 20. It is noted in the resolution that pursuance for happiness is the main purpose of human. Pursuance for happiness and welfare is one of the purposes of this virtual learning too. Therefore, according to Morten Flaten Paulsen, the member of the Board of Virtual learning on the web, the elements of encouraging the students and elements of celebration should not be forgotten during training. FuN (acronym for a web organization Flexible utdanning Norge" translated as "Flexible Education Norway") represents a web Association of flexible education.[4] FuN is a real and successful example of social cluster in Norway. Fun is an acronym for "Fleksibel outdating Norge" which translates to "Flexible Education Norway." FuN organizes 40 Norwegian providers of flexible education; including independent distance education institutions, public universities and colleges, private institutions and training centers for business and industry. Our members provide an extensive

4 FUN. Access through the Internet: http://fleksibelutdanning.no/om-fun/english/

Picture 8: In the photo above: Master's degree student present the project

range of subjects, courses and study programs from primary to university level, and customized courses for professionals. FuN's members offer flexible education at different levels; it could be exclusively online, partly on campus or related to the students' work place. It could be flexible start-ups and all self-paced courses, or common start date and assemblies (online or on campus) along the way. We are working closely with different departments under the Ministry of Education and Research, especially Skills Norway, and we share the responsibility of assessing applications for governmental funding for educational development projects. We also publish an annual report on these projects and the technical and pedagogical development conducted by our members. FuN's members comply with strict rules for quality in flexible education. FuN aims to promote cooperation among its members in matters and initiatives of common interest. Ordinary or consultative membership is being offered, depending on the art of the institution. FuN organizes conferences, discussions, seminars, workshops and field trips for its members and partners.

Research shows that various social indicators are worse in the countries where significant welfare inequality exists. The British researchers Dr. Richard

Wilkinson and Kate Pickett proved that the society much more suffers from various health and social problems in the countries where significant welfare inequality exists: the residents have more overweight, more often suffer from mental illnesses, more homicides and suicides are committed, there are more minor mothers, bigger number of prisoners, a higher level of bulling among the children, high level of drug abuse, etc. compared to the countries with more even distribution of welfare. The researchers first of all link such discovery with an increased level of stress and anxiety in the society that result in parenthood problems, more frequent depression, medication addiction, poorer social life, trust in public institutions and more stress-related diseases. Differences are especially noticeable when the Scandinavian states (where welfare inequality is one of the smallest in the world) are compared with the USA, Germany, Great Britain and even Lithuania. Economist T. Pikkety warns that too high inequality loses its function as promoter of innovations. Moreover, too high inequality of income and assets poses threat to democracy. Assets concentrated in hands of small number of voters cannot anymore ensure democratic political vote. Another problem is that welfare inequality promotes poverty, that is, as more money is concentrated on the top of a pyramid, as less remains on the bottom. The biggest problem of Lithuania: inequality of income and poverty of a large social group. Families facing a long-term poverty often appear in the vicious circle: their members having lost qualification and work skills cannot find work, living in poverty, they are unable to provide their children with higher education, the basics of culture and morals and such adult children face the same problems as their parents. These threats in formation of peculiar zones of poverty in the country, that is, zones where crime, drunkenness, drug addiction and other evil prevail and such evil spreads to other layers of the society. Such poverty can divide the society into semi-closed separate layers living a distinctive life, having a small number of points of contact, often confronting each other. The social security system was reformed over 2017 to 2018, but had a negligible impact on the alleviation of poverty and social exclusion. Health outcomes continue to be one of the lowest in the EU.[5] Do we observe such situation in Lithuania? Our analysis confirms that Lithuania divided in "two Lithuanias" which live different lives. Development of social business in Lithuania could somewhat accelerate **development of social clusters.** Based on cooperation and social partnership business enterprises, science institutions and non-governmental organizations together with science

5 BTI Country Report 2020. Lithuania. Internet Assessment: https://www.bti-project.org/content/en/downloads/reports/country_report_2020_LTU.pdf

organizations can more effectively solve relevant social problems; enable provision of necessary public services for a small fee. Such form of activity would be relevant when developing services of maintenance, care, welfare and employment of children, disabled and elderly people, when developing family and community businesses, which are relevant for the local communities. The opportunities of creation of social clusters should be assessed taking into account the fact to what extent cooperating enterprises and organizations are interested in provision of these services and how much the residents would be ready to pay and to what extent responsible institutions are ready for this. The Ministry of Social Security and Labor notes that now it is a perfect time to assess an opportunity to promote social business in Lithuania more actively and to use the financial measures of the EU for this. Options of addressing the problems interfering with implementation of social business in Lithuania were discussed in the discussion "Social innovations funded by the European social fund – the instrument for reduction of differences of development" organized by the Ministry of Finance in April 2012 as well the examples of good practice were presented, the main trends of the EU support for social innovations in 2014–2020 and the ways of their implementation were discussed. When assessing success of implemented initiates of social business in Lithuania, the potential of their development and plans of the government to promote business development it can be seen that there are big perspectives for social business. Timely made important decisions would result in more opportunities in Lithuania and innovative ways of solving rising social problems and strengthening a capacity of non-governmental sector.

Lithuania

A beehive of non-governmental organizations[6] – a space of work, creation work and cooperation of non-governmental organizations and social innovators. A beehive of non-governmental organizations is designed for creation work and innovative ideas of every initiative person. A beehive of non-governmental organizations creates a pleasant space for acting and sharing. A way of exploitation of this depends on each of us because all of us are responsible for more beautiful world and a socially responsible civil society.

"Beepart" (in Lithuanian "be a part," "bee art" or "working bees") is a public institution which has established art workshops as a center of attraction of the society in Pilaitė (Vilnius). It is a place where unique ideas are born and are implemented.

6 NGO beehive. Access through the Internet netą: http://www.nvoavilys.lt/

In Vilnius, same as in other post-soviet cities, all forms of entertainment and events are concentrated in the central part of the city, mostly in the Old Town. There is an absolute cultural and social calm in a big part of the city or so-called sleeping micro districts and we want to compensate it.

The workshop is oriented to the local society and international projects. Much attention is paid to the improvement of the local social climate; various events involving the local society are organized with help of volunteers. High art schools and their representatives, professional artists and private companies are invited to work together.

Design of the building – Simonas Liūga, co-author of the building – Andrius Ciplijauskas, the author of "Beepart" conception and director of public institution Andrius Ciplijauskas. The building that was designed within two years is characterized by its atypical design, different sense of space and innovative, ecological solutions; it is a kind of exposition house where many innovative solutions can be seen, to "touch" them in reality and to test them. The building constructed of separate modules is independent (except power supply) from the city infrastructure.

Postwoman

The main subject of the initiative – years of press ban in Lithuania. Organizers are planning the project with duration of several years. It is intended to develop a relevant subject in various forms both in Lithuania and abroad because book smuggling is a phenomenon having no analogues in the world. In the first stage, on May 7th, 2017, when Press Restoration Day is celebrated, a Team Field Game "Postwoman" starts. The game is intended to develop a subject of book smuggling in unobtrusive and attractive form of education – footprints of postwoman and later books hidden in the hundred-year-old homestead are searched in accordance with a map given. The local children are also involved in this. Later analogous external game and organizing of educational events, press works and interactive projects are planned to develop the subject of book smuggling as broader as possible. Many people ask why postwoman. During a period of press ban that lasted for 40 years it was impossible to send Lithuanian press by post. Therefore, book smugglers who carried parcels throughout entire Lithuania were also called postal workers. The main actor in the game is Aniutė Kaunietė, the postwoman of Valančius.

Public institution salad bar "Guru" offers persons who are no more addicted to psychoactive substances opportunities of rehabilitation and reintegration in the society and labor marker. "Guru" helps to acquire profession and gain professional experience; it executes professional rehabilitation; motivates people to earn legal income; encourages self-confidence and guarantees employment; enables development of the community of formed drug addicts and promotes various forms of support and self-aid.

" 'Pirmas blynas," Social Pancake restaurant with a culinary and a socially conscious mission. Delicious pancakes are served to you by people with disabilities. The staff will do their absolute best every day in order to ensure that your lunch experience is unforgettable.

"Caritas Works" is a social business. Its main purpose – employment and integration of people experiencing social exclusion (people who have returned from prisons, homeless people, etc.) in the permanent labor market.[7]

Workshop "Creation angle 360°"

Creation angle 360° is a social initiative that develops ideas of sustainable consumption and the secondary design, taking care for development of safer and healthier environment. We carry out educational activity organizing various events and sessions through creativity. The main subject of sessions – the secondary design, that is, we learn to resurrect items for the second life, and we make them from the secondary raw materials or production waste.

Treasures of Old Prussia[8]

Social business of Vilkyškiai and the surrounding communities – collection, drying and realization of medicinal and aromatic plants. The activity is intended to promote employment of young people living in the rural areas. The community of the village organizes the projects that involve young people who are facing troubles, implements educational programs. It is expected to draw attention to unique endangered plants and heritage of Old Prussia through the brand "Treasures of Old Prussia."

Three sisters[9]

It is the project intended to draw women's attention to themselves. The project "Three sisters" operates in the fashion and business center "Heerkaus gallery" in Klaipėda. Here, in the social space for women, is a space for seminars, library and charity store that helps maintain this project and money collected.

7 Caritas Works. Access through the Internet: http://caritasworks.lt/
8 Treasures of Prussia. Access through the Internet: http://prusijoslobiai.lt/
9 Three sisters. Access through the Internet: http://www.trysseserys.lt/

Social taxi[10]

Social taxi is a unique service that helps ensure organizing of traveling of people with movement disability to places of economic and social occupation (work, various organizations, hospital, university, cinema, theatre, store, meeting, etc.).

LABU

Half a year ago, small partnership "LABU" established by three enterprising mothers Akvilė, Jolita and Dovilė has presented to the Lithuanian market a unique sweet. Parents of children eating this sweet are very happy and still cannot believe that a natural product can be so delicious. Therefore, both children and their parents, who also want and like to have healthy snacks, love dried fruit purees "LABU delicacies."[11]

Art after-party. On April 11, 2017, an event "Teacher=Director" of the educational week "Art after-party" took place in Lithuanian National Drama Theatre (LNDT). The cultural educational week "Art after-party" organized for the fourth time is designed to promote cooperation of schools and cultural organizations. During this week, teachers are unable to systematically familiarize educational projects and activities of cultural institutions. The Ministries of Culture, Education, and Science cooperate during implementation of the cultural educational week. Dr. Auksė Petruškevičiūtė, the coordinator of socioeducational projects working for Lithuanian National Drama Theatre, and Paulius Tamolė, an actor and director of the same theatre, conducted the workshop "Teacher=Director" for educators who have gathered in the theatre. Teachers of different subjects from various places of Lithuania – Vilnius, Kėdainiai, Marijampolė, Kazlų Rūda, Širvintos, Vilnius district – have gathered in the theater. The main purpose of the workshop is to help educators develop competencies that are necessary for enrichment of organizing of educational process. While working in groups and individually, teachers were developing their creativity, ability to work in the team, to lead a group, to have learned how to develop imagination and innovative thinking and to find sources of creativity in the daily life. Paulius Tamolė has revealed to the participants of the workshop

10 Social taxi. Access through the Internet: http://socialinistaksi.lt/

11 Enterprising mothers present healthy treats. Access through the Internet:: http://www.mamyciuklubas.lt/moteru-klubas/verslios-mamos-jolita-dovile-ir-akvile-pristato-sveikus-skanestus-labu-48540/

the methods of acting and directorial creativity that can be also adapted in educational work. The participants of the workshop have also familiarized other opportunities of enrichment of educational processes offered by LNDT.[12] This project is already continuous. However, it does not have yet the main accents of social business (purchases and sales). Notwithstanding this, more and more projects directly associated with social business occur in Lithuania.

LECH LECHA **Lech Lecha**

The initiative of social business LECH LECHA Design is a fashion brand of new generation, which talks about serious things in a positive, happy and colorful way. With help of clothing design, we promote the society to draw attention to human: to both human suffering poverty, homelessness and social exclusion and the one who is successful, that is, who is indifferent to these problems. We create unique clothing and accessories providing both clothes and people with the second chance. Instead of fabrics, we use the existing textile products. We remake because we care about the nature and wish to create a meaningful fashion. People of social exclusion help create clothing and accessories of unique design during art therapies. We involve in creative processes with people experiencing homelessness and poverty and those who are recovering from addictions. We can see more sense in providing people and fabrics with the second chance than in simply creating a short-term new thing. We are a team that cares not only about fashion!

Cool Ūkis **Cool Farm**

The city community gait connects willingness to cultivate own harvest and to share land and experience. Young townspeople strongly wish to cultivate their own harvest, to acknowledge, communicate and take care for the environment. Whereas older people have land, they no longer use that or they are unable to use it. However, they would like to share harvest and knowledge. The purpose of social business – to connect these wishes and to take care for their implementation.

"Rafaelis" established by Marija Mendelė-Leliugienė has implemented the project "Solution for you." It has developed the methodology – development of skills necessary for enterprise using art therapy for unemployed

12 Art after-party. Access through the Internet: http://www.teatras.lt/. . ./teatre_-_menu_duzgiu_renginys_moky. . ./; http://www.upc.smm.lt/naujienos/duzges/ivadinis2017.php

women or those who wish to return to the labor market after maternity leave. Often such women have insufficient qualification, communication skills or avoid the society. In order to return women to the labor market, the program that has already been applied in seven countries of the world was created. It was bought as a service by the United Nations (UN) for development of enterprise of homeless people. Good practice of this program – 58 of 60 homeless people have drawn up business plans, two of them were evaluated as perfect. Another example of success of the program – one homeless woman who prepared the project managed to find a job before its end. The project of public institution "Rafaelis" that cooperates with the Ministry of Educations and Science of the Republic of Lithuania is creation and integration of nontraditional educational methods in general education, after-school activities and art and music therapy. Research was performed with 1500 children and 150 specialists were trained. The project took place in five regions. Its result: the methodology with visual material was issued.

Children's temporary shelter "Turn yourself to children"[13] has implemented the project "Guaranteed start" funded by the EU. The project is designed for occupational preparation and employment of young people. Support for funding of new youth home was received. Currently fostered young people and children who have left foster home are unemployed but ready to acquire profession live in a newly opened home. Young people are advised, taught English language, computer literacy and initial occupational skills. Thus, young people participating in the project acquire professions of cook, waiter, bartender and house cleaner. A meaningful project helps undisciplined young people with behaviors problems find job, integrate themselves in the labor market and meet the employer's requirements. Children's temporary shelter "Turn yourself to children" has long years of experience of work with Norwegian "Lions club." Norwegian citizens, seeking a positive socialization of children, have established non-governmental organization "Turn yourself" in their country, advised employees of the shelter on the issues of positive socialization and psychological support, which are especially important for socially vulnerable children. The partnership has developed from acquaintance of Giedrė Kvieskienė (Lithuania) and Einar Kjonar (Norway) and their agreement and continues already for 16 years. Twice a year the Norwegian club provides charity on demand. In 2014, a concert of support was organized already for the ninth time. Moral and financial support further encourages continuation of activities.

13 Turn yourself to children. Access through the Internet: http://atsigrezk.org/

All activity of **NGO confederation for children**[14] is based on the development of social partnership by nurturing the child's well-being. During the years of its activity NGO confederation for children has not only gathered the strongest NGOs for united activity but also started active activity both in national and regional levels regarding implementation of fundamental child's rights in Lithuania. The biggest and the most successful projects based on a social partnership in the field of child's welfare of NGO confederation for children are presented hereinafter.

In 2011, when European Men's Basketball Championship "EuroBasket 2011" was organized, more than 1100 children from member organizations of NGO confederation for children (foster homes, centers of social services, day centers, etc.) watched U18 Star match in Kaunas. The children were allowed to see a show of the best young European basketball stars. These are practical skills of formation of sports life attitude, experience and positive emotions. Young basketball players show good example to children how hard work and seeking of established goal allows achieving perfect results. Since the children attending day centers and living in foster homes are not only passionate basketball fans but also like to play basketball, they waited eagerly for the match which got a lot of positive emotions and new impressions. Almost 300 children from different member organizations of NGO confederation for children have opportunity to watch free of charge the match of the championship that was held in Alytus. NGO confederation for children has signed the cooperation agreement with the organizational committee of the European Men's Basketball Championship of 2011 and Lithuanian Basketball Federation whereby social projects and initiatives that encourage paying attention to social problems, education of civil consciousness, promotion of interest in sport and a healthy lifestyle are organized. As result of cooperation, a social campaign "Stop violence against children!" was organized in 2010. Famous basketball players posed for photos with children asking the society to contribute in solving problems of violence against children. In 2011 in cooperation with social partner's quiz, a relevant basketball issue was organized. Severely disabled people who took part in the quiz were invited to the matches of the European Men's Basketball Championship that were held in all cities.

14 NvO Vaikams Konfederacija. Access through the Internet: https://www.facebook.com/nvovaikamskonfederacija/

Initiative of NGO confederation for children can be considered an example of good practice: Annual International Conference is organized since 2005 together with partners. The Conference is designed for analysis of implementation of the UN Convention on the Rights of the Child in Lithuania, for the members of NGO confederation for children and the organizations operating in the field of child's welfare, national and international partners. The Conference is a part of monitoring of the situation of child's welfare. It is intended for exchange of information, to promote mutual trust and partnership. The representatives of various fields share their experience of social partnership and cluster ship, analyze processes that promote mechanisms of socioeconomics and discuss about further opportunities of partnership, innovations in establishing a safe childhood and creation of family friendly environment and development of harmonious and sustainable society. Speakers try to look at development of child and family friendly environment, social partnership in their presentations of the international conference through a prism of positive socialization. The representatives from Lithuania, Latvia, Great Britain, Germany, Azerbaijan, the Slovak Republic, USA, Canada and other countries took part in the Conference. Presentations were made by the representatives of the national institutions, the representatives from science, educational institutions and the representatives of responsible business and political parties.

The Christmas event for children "Children constellation" is another example of partnership between NGO, private and public sectors and one of the best-known initiatives of the NGO confederation for children that is organized every year since 2005. Nomination "Best children's friend" are held every year and in this ceremony NGO elect most dedicated people in children welfare policy from public, private, NGO and Science sectors, become one of prestigious nomination in Lithuania. The event gathers the children of the member organizations of NGO confederation for children. This event is the only opportunity to celebrate Merry Christmas by majority of these children. Every year more and more children participate in the event: in 2011 a number of children compared to 2010 has doubled – 1200 children (only slightly more than 600 children took part in the event in 2010). It is an excellent initiative that unites NGO, state and private sectors and promotes children's welfare. It is important to note that in 2011, the Chairman of the Seimas of the Republic of Lithuania was the sponsor of "Children constellation" and "Danske Bank" was the patron of the event. Every year the children get Christmas presents and eat Christmas cake. Linas Kleiza who was awarded nomination "The children's friend" and the Charity and Support Foundation established by him sent a cake to the children in 2010.

Successful partnerships of **Trakai Education Center** that promote child's welfare in establishment of training/education institution for the street children, for example, "Sodžiaus meistrai" in Rūdiškės, Trakai district, as well in implementation of the project of the European Social Fund (ESF) initiated by the Ministry of Education and Health of the Republic of Lithuania designed for improvement of education of students with special educational needs. By the first project, the school every year educates approximately 12 students who are threatened by blind-alley future or maybe even imprisonment. In the second case – all students with special education needs successfully complete academic year and their behaviors and feelings improve. The majority of these students live in foster homes. The budgets of the projects consisted of contribution of private sponsors, support through the projects for development of material and intellectual welfare of the institution and support of the government.

Several social companies based on the principles of social business and social clustering also operate in **Lithuania**. One of the examples, "Grikė,"[15] a company that manufactures buckwheat hull cushions operates in Vilnius. It is led by disabled people. Most of the employees of the company also are disabled people. The company operates in accordance with democratic principles. Unlike in the major corporations where a meeting of few people decides regarding a fate of thousands of people, its members have the right to express their will during the process of decision-making. Another inspiring example is an already presented coffee shop "Mano Guru" that operates in the capital already eight years. Addicted people undergoing treatment work in a stylish restaurant-salad bar operating in the central street of the city. Nevertheless, it is obvious that the establishment has difficulties of competing with other restaurants established in the same street. However, with a little aid of the state it can survive and create welfare for the society – after having received a little bit of financial injections from the state the coffee shop integrates vulnerable people to the society, teaches them working skills. Only business or only the state could not achieve such results. Sceptics may say that non-governmental organizations already operate in Lithuania because public groups representing individual sectors can be found in every municipality. However, practically such groups have no real impact on the Board. Such organizations can be identified as sham accompaniments to the City Boards, public organizations, subjects of socioeconomics would be more influential, and then more social and economic problems would have been solved already. The authorities should

---

15 Grikė. Access through the Internet: http://www.grike.lt/Kontaktai/

be more determined and rely much more on Lithuanian people. They should only be provided with appropriate tools. Authorities should check how money is being spent and support people in case of extended funding if people execute their plans and effectively use support provided to them.

All success cases are characterized by social partnership, network cooperation and trust. Trust guarantees a quality of relations. Therefore, it is very important not only to cooperate with the family, the job team, but also to participate in the public and civil activity. Communication and partnerships are the most important activities of the human, which young people can learn in educational institutions. Partnership and trust encourage creation work and help to make decisions. The greater the cooperation is, the greater the social capital is generated, and social networks are developed.

Discussion

In this monograph, there is a result of long-term cooperation between Vytautas Magnus University and Indiana University, Bloomington[16], USA, Heidelberg University, Germany, other social partners from Norwegian, Finland, Denmark, other countries research project, which was managing from 2000 till 2018. All our partners have long experience in the field of social partnership in educational institutions.

Indiana University School of Education is one of the biggest educational schools in the USA, which focuses on educating teachers, innovative education, for habits model and smart cities innovations, as well as scholars and educational evaluators within the network of several units operating within IU. School of Education researchers are leading actors in Civic Education Networks and pioneers of integrated educational innovation in Secondary Schools. The long-term partnership between Indiana University and Lithuanian Educational University has encouraged us to create joint publications[17].

This monograph project presents a universal networking design in research, private, public sectors and NGO's (Kvieskienė, Kvieska, 2012). Social clustering imperative shows a new meaning by developing a new perspective on how to empower representatives from various backgrounds: universities, business, authorities and NGOs. However, a universal design climate and culture cannot

16 School of Education. Access through the Internet: http://education.indiana.edu/

17 Kvieskienė, G., Masson, T. C. Integrating Civic Education and Social Work: The Social Educator Master's Degree Program at Vilnius Pedagogical University. In: Civic Education in Teacher Education, Volume 3, The Social Studies Development Center

be cultivated until the social cluster practices are created. Lastly, our efforts in finding successful practices in social clustering stakeholders creating sharing culture is the beginning of a complete educational transformation, in order to structure the new culture in a universal design.

Social and economic changes taking place all around the world, including Lithuania, lead to change of abilities and competences that are necessary in the labor market and a social environment. Skills of cooperation and social communication become more and more important. Tendency and ability to communicate effectively in the all levels of the social groups and society, improving trust and giving us better social emotional health, to making our life long projects is described by a term "social capital."

Influence of a human capital on economic growth and innovations has been proven long time ago.

Our days performed empirical macroeconomic and social research revealed that social capital plays an important role not only when economic growing but when we can create sustainable interfaces between social capital and health and education innovations for welfare communities. By strengthening development of innovations, it may promote economic growth. Moreover, its interfaces with various social aspects, that is, human health, a level of criminality and a number of suicides, were revealed. Data of the research presented confirm a fact that a level of social capital and social competences in Lithuania is low compared to other European countries. This, in turn, affects efficiency of economic activity and has negative social consequences. Two main means of public policy – activation of civic society and improvement of access to education, its flexibility and quality, increase of its variety and accessibility – are recommended to improve a situation.

---

of Indiana University in Association with Civitas: An International Civic Education Exchange Program. Edited by Gregory E. Hamot, John J. Patrick and Robert S. Leming: Indiana University Social Studies Development Center, USA, 2004. ISBN 0-941339-28-9, p. 189–215; (2). Kvieskienė G., Metcalf, K., Nixson, W.A, Patrick, J.J., Vontz, T.S. Implementation of Project Citizen in Indiana, Latvia and Lithuania, in Project Citizen in Indiana, Latvia, and Lithuania, USA: ERIC, 2000, p. 103–125. ISBN 0-941337-25-4; (3). Kvieskienė G., Žadeikaitė L., Patrick J. J. Constitutional Democracy in Lithuania//Civic Education in Teacher Education. Volume 4. The Social Studies Development Center of Indiana University in Association with Civitas: An International Civic Education Exchange Program. Edited by Gregory E. Hamot, John J. Patrick and Gregory E. Hamot: Indiana University Social Studies Development Center. 2005. 180–217 p. ISBN 0-9773916-0-4

The focus in this monograph is on leveraging social capital through social partnership between teachers, school leaders, NGO, communities and private sector in the area of curriculum change and instructional excellence. Influence of social partnership on development of novelties, social innovations and investment. Partnership of several sectors always initiates occurrence of new ideas, jobs, improvement of social-ecological environment and quality of life and a better and more efficient way of social solutions. Based on social partnership social networks, other umbrella and network organizations increase intersectional trust form.

Tasks of socioeconomics – to change traditional way of thinking about role of schools, to seek optimization of social economy and promote entrepreneurship and to develop economy of services together with an organized civic society based on phenomena of a social partnership and networking.

In most countries, schools are not encouraged to take the lead in initiating social partnerships and to organize activities so that teachers can systematically collaborate not only with one another, but also with the local community or business to implement smart education programs and initiate transformations. In most cases, teachers are still relatively isolated, rarely seeing another colleague, unless they meet in hallways, in rare integration lessons, or in teachers' rooms. But in their classrooms, working directly with children has a huge impact on children. They work behind closed doors for maybe 4 hours directly, up to 160 students a day, literally hundreds of teacher-student communication. In doing so, they have almost unlimited autonomy in the classroom, the pace of teaching, what they teach, how they teach and how they value their students. The requirement for teachers to cooperate more with each other, with parents and with the community means that they need to find time and organize more lessons outside the school.

A quality social partnership between public, private and NGO sectors improves socioecological environment and a quality of human life, develops social capital and participation of civic organizations, improves life of communities and promotes innovations and investment.

Trust is essential for a social partnership and network cooperation. It ensures a quality of relations. Therefore, it is very important not only to communicate in the family and work team but also to participate in public, civic activity. Communication, partnership are the most important human activities. Partnership and trust promote creation and help make decisions faster. We have measured the cooperation is, through indicators of composition the social capital and spatial social networks developed.

Social partnership and social clustering promote social innovations and small- and medium-sized business, help solve chronic social problems and cultivate intersectoral trust, thus increasing social capital. In the development of social clustering, social business is associated with evolution of socioeconomics and tradition of the country. The principles of an effect of social clustering depend not only on socioeconomic situation of the country, but also on a fact, which organizations are strong: NGOs, charity organizations or semi-public institutions. When a strong welfare state is absent, a social business is often identified as social services and takes over a part of functions of public sector. However, goals of a social business should be considered more pragmatically in order it would not take over but would be able to solve or at least more effectively deal with social problems. Otherwise, no change occurs and status quo remains when a problem is simply being other hands. Social clustering helps integrate the existing resources, generate additional resources and income, and solve multi-layered social human problems by providing them with a clear valuable basis when solving relevant problems (e.g., pollution and inadequate wages). Therefore, when preparing and studying these modules we were striving for social changes, were searching for problems that require new solutions because public services only satisfy a need but in principle leave a problem unsolved. Most of modules analyzed from other countries naturally require other business model and a broader perception of solutions for social problems.

Results of empirical research, indexes of public welfare and happiness and analysis of good experience clearly demonstrate a relation of social competences and social capital with economic and social welfare. The authors conclude that for school leadership in smart education a new leadership model is required when attempting to overcome the inertia in schools toward school reform. One of the scenarios there is move from leader-initiated transactions to research and data driven, teacher- and student-initiated project, which is initiated together with local community, NGO and entrepreneurship.

Authors argue that school leadership needs to move away from a "leader-centric" organizational framework especially in a digital networking culture of change and innovation.

Social competences play a more and more important role in the developing knowledge-based society – knowledge personnel currently spend most of their working time searching for information, assessing it and communicating. Innovations are very important for knowledge-based economy, especially ability to intercept and improve them. Innovations, especially ability to take them over and improve are very important for knowledge-based economy.

In order to use possibilities of all employees to develop innovations in their work, advanced companies implement modern forms of work organizing, that is, high-performance work systems. According to data of international comparative research, Lithuania lacks both social capital and social competences. The country is also characterized by a significant imbalance between qualification of employees, their competences and demands of the labor market. Due to interfaces of social capital and social skills with economic growth, innovations and social welfare, their lack negatively influences the said factors. A level of social capital and competences must be raised to achieve a high level of economic and social welfare. This can be best achieved with help of two main means of public policy – by promoting activity of the society (participation of people in the process of decision-making and in activity of non-governmental organizations and creation of environment suitable for this) and by improving a quality of education, by increasing its diversity and accessibility (by implementing and developing modern forms of teaching in the system of both formal and informal education).

Annexes

The European Parliament, the Council and the Commission solemnly proclaim the following text as the European Pillar of Social Rights.

European Pillar of Social Rights[1]

The completion of the European single market in the last decades has been accompanied by the development of a solid social acquis which has resulted in progress in the freedom of movement, living and working conditions, equality between women and men, health and safety at work, social protection and education and training. The introduction of the euro has provided the Union with a stable common currency shared by 340 million citizens in nineteen Member States, facilitating their daily lives and protecting them against financial instability. The Union has also enlarged significantly, increasing economic opportunities and promoting social progress across the continent.

Labor markets and societies are evolving quickly, with new opportunities and new challenges arising from globalisation, the digital revolution, changing work patterns and societal and demographic developments. Challenges, such as significant inequality, long-term and youth unemployment or intergenerational solidarity, are often similar across Member States although in varying degrees.

Europe has shown its resolve to overcome the financial and economic crisis, and as a result of determined action, the Union economy is now more stable, with employment levels at an unprecedented high and a steady fall in unemployment. However, the social consequences of the crisis have been far-reaching – from youth and long-term unemployment to the risk of poverty – and addressing those consequences remains an urgent priority.

To a large extent, the employment and social challenges facing Europe are a result of relatively modest growth, which is rooted in untapped potential in terms of participation in employment and productivity. Economic and social progress are intertwined, and the establishment of a European Pillar of Social Rights should be part of wider efforts to build a more inclusive and sustainable

1 The European Parliament, the Council and the Commission solemnly proclaim the following text as the European Pillar of Social Rights. EUROPEAN PILLAR OF SOCIAL RIGHTS. Access through the internet: http://eur-lex.europa.eu/legal-content/LT/TXT/?uri=COM:2017:251:FIN

growth model by improving Europe's competitiveness and making it a better place to invest, create jobs and foster social cohesion.

The aim of the European Pillar of Social Rights is to serve as a guide toward efficient employment and social outcomes when responding to current and future challenges, which are directly aimed at fulfilling people's essential needs and ensuring better enactment and implementation of social rights.

A stronger focus on employment and social performance is particularly important to increase resilience and deepen the Economic and Monetary Union. For this reason, the European Pillar of Social Rights is primarily conceived for the euro area, but it is applicable to all Member States that wish to be part of it.

The European Pillar of Social Rights expresses principles and rights essential for fair and well-functioning labor markets and welfare systems in 21st century Europe. It reaffirms some of the rights already present in the Union acquis. It adds new principles which address the challenges arising from societal, technological and economic developments.

The principles enshrined in the European Pillar of Social Rights concern Union citizens and third country nationals with legal residence. Where a principle refers to workers, it concerns all persons in employment, regardless of their employment status, modality and duration.

The European Pillar of Social Rights shall not prevent Member States or their social partners from establishing more ambitious social standards. In particular, nothing in the European Pillar of Social Rights shall be interpreted as restricting or adversely affecting rights and principles as recognized, in their respective fields of application, by Union law or international law and by international agreements to which the Union or all the Member States are party, including the. European Social Charter signed at Turin on 18 October 1961 and the relevant Conventions and Recommendations of the International Labour Organisation.

Delivering on the European Pillar of Social Rights is a shared commitment and responsibility between the Union, its Member States and the social partners. The principles and rights set by the European Pillar of Social Rights should be implemented at both Union level and Member State level within their respective competences and in accordance with the principle of subsidiarity.

At Union level, the European Pillar of Social Rights does not entail an extension of the Union's powers as defined by the Treaties. It should be implemented within the limits of those powers.

At Member State level, the Pillar respects the diversity of the cultures and traditions of the peoples of Europe, as well as the national identities of the Member States and the organisation of their public authorities at national, regional and local levels. In particular, the establishment of the Pillar does not affect the right of Member States to define the fundamental principles of their social security systems and should not affect the financial equilibrium thereof.

Social dialogue plays a central role in reinforcing social rights and enhancing sustainable and inclusive growth. Social partners at all levels have a crucial role to play in pursuing and implementing the European Pillar of Social Rights, in accordance with their autonomy and the right to collective action.

European Pillar of Social Rights

Chapter I: Equal Opportunities and Access to the Labor Market

1. Education, Training and Life-Long Learning

Everyone has the right to quality and inclusive education, training and life-long learning in order to maintain and acquire skills that enable them to participate fully in society and manage successfully transitions in the labor market.

2. Gender Equality

a. Equality of treatment and opportunities between women and men must be ensured and fostered in all areas, including regarding participation in the labor market, terms and conditions of employment and career progression.
b. Women and men have the right to equal pay for work of equal value.

3. Equal Opportunities

Regardless of gender, racial or ethnic origin, religion or belief, disability, age or sexual orientation, everyone has the right to equal treatment and opportunities regarding employment, social protection, education and access to goods and services available to the public. Equal opportunities of under-represented groups shall be fostered.

4. Active Support to Employment

a. Everyone has the right to timely and tailor-made assistance to improve employment or self-employment prospects. This includes the right to receive support for job search, training and re-qualification. Everyone has the right to transfer social protection and training entitlements during professional transitions.
b. Young people have the right to continued education, apprenticeship, traineeship or a job offer of good standing within 4 months of becoming unemployed or leaving education.

c. People unemployed have the right to personalised, continuous and consistent support. The long-term unemployed have the right to an in-depth individual assessment at the latest at 18 months of unemployment.

Chapter II: Fair Working Conditions

5. Secure and Adaptable Employment

a. Regardless of the type and duration of the employment relationship, workers have the right to fair and equal treatment regarding working conditions, access to social protection and training. The transition toward open-ended forms of employment shall be fostered.
b. In accordance with legislation and collective agreements, the necessary flexibility for employers to adapt swiftly to changes in the economic context shall be ensured.
c. Innovative forms of work that ensure quality working conditions shall be fostered. Entrepreneurship and self-employment shall be encouraged. Occupational mobility shall be facilitated.
d. Employment relationships that lead to precarious working conditions shall be prevented, including by prohibiting abuse of atypical contracts. Any probation period should be of reasonable duration.

6. Wages

a. Workers have the right to fair wages that provide for a decent standard of living.
b. Adequate minimum wages shall be ensured, in a way that provide for the satisfaction of the needs of the worker and his/her family in the light of national economic and social conditions, while safeguarding access to employment and incentives to seek work. In-work poverty shall be prevented.
c. All wages shall be set in a transparent and predictable way according to national practices and respecting the autonomy of the social partners.

7. Information about Employment Conditions and Protection in Case of Dismissals

a. Workers have the right to be informed in writing at the start of employment about their rights and obligations resulting from the employment relationship, including on probation period.

b. Prior to any dismissal, workers have the right to be informed of the reasons and be granted a reasonable period of notice. They have the right to access to effective and impartial dispute resolution and, in case of unjustified dismissal, a right to redress, including adequate compensation.

8. Social Dialogue and Involvement of Workers

a. The social partners shall be consulted on the design and implementation of economic, employment and social policies according to national practices. They shall be encouraged to negotiate and conclude collective agreements in matters relevant to them, while respecting their autonomy and the right to collective action. Where appropriate, agreements concluded between the social partners shall be implemented at the level of the Union and its member states.
b. Workers or their representatives have the right to be informed and consulted in good time on matters relevant to them, in particular on the transfer, restructuring and merger of undertakings and on collective redundancies.
c. Support for increased capacity of social partners to promote social dialogue shall be encouraged.

9. Work-Life Balance

Parents and people with caring responsibilities have the right to suitable leave, flexible working arrangements and access to care services. Women and men shall have equal access to special leaves of absence in order to fulfill their caring responsibilities and be encouraged to use them in a balanced way.

10. Healthy, Safe and Well-Adapted Work Environment and Data Protection

a. Workers have the right to a high level of protection of their health and safety at work.
b. Workers have the right to a working environment adapted to their professional needs and which enables them to prolong their participation in the labor market.
c. Workers have the right to have their personal data protected in the employment context.

Chapter III: Social Protection and Inclusion

11. Childcare and Support to Children

a. Children have the right to affordable early childhood education and care of good quality.
b. Children have the right to protection from poverty. Children from disadvantaged backgrounds have the right to specific measures to enhance equal opportunities.

12. Social Protection

Regardless of the type and duration of their employment relationship, workers, and, under comparable conditions, the self-employed, have the right to adequate social protection.

13. Unemployment Benefits

The unemployed have the right to adequate activation support from public employment services to (re)integrate in the labor market and adequate unemployment benefits of reasonable duration, in line with their contributions and national eligibility rules. Such benefits shall not constitute a disincentive for a quick return to employment.

14. Minimum Income

Everyone lacking sufficient resources has the right to adequate minimum income benefits ensuring a life in dignity at all stages of life, and effective access to enabling goods and services. For those who can work, minimum income benefits should be combined with incentives to (re)integrate into the labor market.

5. Old Age Income and Pensions

a. Workers and the self-employed in retirement have the right to a pension commensurate to their contributions and ensuring an adequate income. Women and men shall have equal opportunities to acquire pension rights.
b. Everyone in old age has the right to resources that ensure living in dignity.

16. Health Care

Everyone has the right to timely access to affordable, preventive and curative health care of good quality.

17. Inclusion of People with Disabilities

People with disabilities have the right to income support that ensures living in dignity, services that enable them to participate in the labor market and in society, and a work environment adapted to their needs.

18. Long-Term Care

Everyone has the right to affordable long-term care services of good quality, in particular home-care and community-based services.

19. Housing and Assistance for the Homeless

a. Access to social housing or housing assistance of good quality shall be provided for those in need.
b. Vulnerable people have the right to appropriate assistance and protection against forced eviction.
c. Adequate shelter and services shall be provided to the homeless in order to promote their social inclusion.

20. Access to Essential Services

Everyone has the right to access essential services of good quality, including water, sanitation, energy, transport, financial services and digital communications. Support for access to such services shall be available for those in need.

List of Figures

List of Pictures

List of Tables

References

Ashoka. https://www.ashoka.org (07.12.2020).

Bergold, J. & Thomas, S. (2012). Participatory Research Methods: A Methodological Approach in Motion. FQS. Forum: Qualitative Social Research., No. 1, Art. 30, January. http://www.qualitative-research.net/index.php/fqs/article/view/1801/3334 (07.12.2020).

Beržinskas, A. (2015). Application of Gamification in the Process of Corporate Idea Generation. Master's thesis. Vilnius: VDU.

BIS (2013a). Smart Cities Background Paper. London: Department for Business Innovation and Skills.

BIS (2013b). The Smart City Market: Opportunities for the UK. London: Department for Business Innovation and Skills.

Bitinas, B. (2004). Socialinis auklėjimas (Social education). In: Hodegetika (Hodegetics). Vilnius: Kronta, pp. 155–164.

Borzaga, C. & Galera, G. (2016). A map of social enterprises and their eco-system in Europe. file:///C:/Users/admin/Downloads/Synthesis%20report%20FINAL.pdf (07.12.2020).

Brinkley, I. (2013). Flexibility or Insecurity? Exploring the Rise in Zero Hours Contracts. London: The Work Foundation.

Cardone, O. [orocardone] (2015). Analysis and critical evaluation of the ethical standards of the UK's No.1 Sports Retailer. Posted on December 5. International Management. https://ocinternationalmanagement.wordpress.com/2015/08/06/%ef%bf%bcanalysis-and-critical-evaluation-of-the-ethical-standards-of-the-uks-no-1-sports-retailer/.

Caritas Works. http://caritasworks.lt/ (07.12.2020).

Carlsburg G.-B. v. (Hrsg./ed.) (2016). Strategien der Lehrerbildung. Zur Steigerung von Lehrkompetenzen und Unterrichtsqualität. Strategies for Teacher Training. Concepts for ImprovingSkills and Quality of Teaching (= Baltische Studien zur Erziehungs- und Sozialwissenschaft. Band 31). Frankfurt: Peter Lang Edition.

Carlsburg, G.-B. v. (Hrsg./ed.) (2017). Denk- und Lernkulturen im wissenschaftlichen Dialog. Cultures of Thinking and Learning in the Scientific Discourse (= Baltische Studien zur Erziehungs- und Sozialwissenschaft. Band 32). Frankfurt : Peter Lang Edition

Carlsburg, G.-B. v. (Hrsg./ed.) (2019). Transkulturelle Perspektiven in der Bildung. Transcultural Perspectives in Education (= Baltische Studien zur Erziehungs- und Sozialwissenschaft. Baltic Studies in Educational and Social Sciences. Band 34). Berlin: Peter Lang Edition.

Carlsburg, G.-B. v. & Wehr, H. (Hrsg.) (2018). Kooperatives Lehren und Lernen lernen (= Pädagogik und Psychologie). 5. Auflage. Augsburg: Brigg.

Carlsburg, G.-B. v., Dammer, K.-H. & Wehr, H. (Hrsg.) (2019). „Hätte ich doch nicht weggeschaut!" Zivilcourage früher und heute (= Pädagogik). 3. Auflage. Augsburg: Brigg.

Carlsburg, G.-B. v., Juceviciene, P. & Merkys, G. (2005). Learning and Development for Innovation, Networking and Cohesion (=Baltische Studien zur Erziehungs- und Sozialwissenschaft. Band 12). Frankfurt: Peter Lang.

Celiešiene., E., Kvieskienė, G., Kaušylienė, A. & Trukanė, I. (2019). Play for children with disabilities: the users' needs in the Lithuanian context. In: Mara Westling Allodi & Tamara Zappaterra (Eds.). Play for Children with Disability: Case Studies. Case Studies from Sweden, Finland and Lithuania. https://content.sciendo.com/downloadpdf/title/527467PDF file (07.12.2020).

Civitas International Programs. http://www.civiced.org/programs/civitas (07.12.2020).

Communication from the Commission to the Council, the European Parliament, the European Economic and Social Committee and the Committee of the Regions. A lead European market initiative {SEK(2007) 1729} {SEK(2007) 1730}. http://eur-lex.europa.eu/legal-content/LT/TXT/?uri=CELEX:52007DC0860 (07.12.2020).

Crane, A. & Matten, D., (2007). Business Ethics: Managing Corporate Citizenship and Sustainability in the Age of Globalization. Oxford: University Press.

Crane, A. & Ruebottom, T. (2011). Stakeholder theory and social identity: rethinking stakeholder identification. Journal of Business Ethics, 102, 1, pp. 77–87.

Creative Approaches to Health and Social Care Education: Knowing Me, Understanding You? (2010). Basingstoke; New York (N.Y.): Palgrave Macmillan.

De Los Ríos-Carmenado, I. & Ortuño, M. (2016). María Rivera. Private-Public Partnership as a Tool to Promote Entrepreneurship for Sustainable Development: WWP Torrearte Experience. http://www.mdpi.com/2071-1050/8/3/199/htm (07.12.2020).

Degėsys, L. (2009). Theory of Education as the Deconstruction and Interpretation of Social Reality. Žmogus ir žodis : mokslo darbai. Flosofija. Vilnius : Vilniaus pedagoginis universitetas., 2009, T. 11, nr. 4, pp. 41–46.

Dėl Lietuvos Respublikos Vyriausybės 2016 metų veiklos ataskaitos pateikimo Lietuvos Respublikos Seimui. 2017 m. kovo 29 d. Nr. 234. https://www.e-tar.lt/portal/lt/legalActPrint?documentId=8d5c7320161511e79800e8266c1e5d1b. (08.12.2020).

Duckenfield, T (2014). What People Want from Their Cities, Connected Cities 2014, London: Steer Davies Gleave.

European Commission – Directorate-General for Employment, Social Affairs and Inclusion Directorate E Unit E1. Authors: Carlo Borzaga and Giulia Galera. Contact: Risto Raivio (2016). Social Enterprises and their Eco-systems: Developments in Europe. Luxembourg: Publications Office of the European Union. http://base.socioeco.org/docs/ke-01-16-853-en-n_-_report.pdf (08.12.2020).

European Foundation for the Improvement of Living and Working Conditions (2017). Luxembourg: Office for Official Publications of the European Communities. Pasaulio ekonomikos forumas. https://www.weforum.org/press/2017/03/world-economic-forum-europeaninvestment-bank-urge-eu-finance-ministers-to-make-inclusive-growthtop-priority (08.12.2020).

European Family Therapy (2020). What is family therapy? Posted on November 23. http://www.europeanfamilytherapy.eu/what-is-family-therapy (08.12.2020).

Europos Sąjungos Tarybos rezoliucija dėl atnaujintos Europos suaugusiųjų mokymosi darbotvarkės 2011/C 372/01. European Council Resolution on a renewed European agenda for adult. Learning. http://eur-lex.europa.eu/legal-content/LT/TXT/HTML/?uri=CELEX:32011G1220(01)&from=LT (08.12.2020).

Frick, T.W. (2012). The Theory of Totally Integrated Education: TIE. A Monograph in Four Parts. Blooming, Indiana: Department of Instructional Systems Technology School of Education, Indiana University. ISBN 094978401X. http://educology.indiana.edu/Frick/TIEtheory.pdf (07.12.2020).

Frick, T.W. (2017). The Theory of Totally Integrated Education: TIE, Indiana University Bloomington frick@indiana.edu (last revised June 29, 2017). https://educology.indiana.edu/Frick/TIEtheory.pdf (07.12.2020).

FuN (2008/20). http://fleksibelutdanning.no/om-fun/english/. (07.12.2020).

Gamification of learning. https://www.linkedin.com/learning/gamification-of-learning?trk=slideshare_sv_learning (07.12.2020).

Government of Lithuania (2017). Dėl Psichikos sveikatos strategijos įgyvendinimo 2007–2016 metų. Vilnius: Lietuvos Respublikos Vyriausybės.

Gudauskas, R. & Kušylienė, A. (2011). Žinių vadyba ir edukacinės inovacijos. Vilnius: Lietuvos edukologijos universitetas.

Gulbinas, R. (2014). Socialinio pedagogo informacinėmis ir komunikacinėmis technologijomis praturtintos edukacinės aplinkos pedagoginis pagrindimas, daktaro disertacija: socialiniai mokslai, edukologija (07 S)/Rokas Gulbinas. Lietuvos edukologijos universitetas. Vilnius: Edukolologija, 229 p.

Holder, G. den (2016) How to Start a Social Business [Guide]. Sept. 9, 2016. https://www.business2community.com/social-business/start-social-business-guide-01651630 (07.12.2020).

How Poverty affects the Brain. http://www.newsweek.com/2016/09/02/how-poverty-affects-brains-493239.html (07.12.2020).

Katinaitė-Lodh, R. (2013). Švietimo prieinamumas nuteistiems laisvės atėmimo bausme asmenims: aukštojo mokslo aspektas: daktaro disertacija: socialiniai mokslai, edukologija (07 S)/Renata Katinaitė-Lodh. Lietuvos edukologijos universitetas. Vilnius: Edukologija, 229 p.

Kazakevičiūtė, I. (2017): What the USA could learn from the globally best Norwegian democracy. Posted on 24 vasario. Ko JAV gali pasimokyti iš Norvegijos geriausios pasaulio demokratijos – Skandinavijos naujienų portalas (skandinavija.today). https://skandinavija.today/naujienos/pasaulio-naujienos/ko-jav-gali-pasimokyti-norvegijos-geriausios-pasaulio-demokratijos/ (08.12.2020)

Komisijos komunikatas Tarybai, Europos Parlamentui, Europos ekonomikos ir socialinių reikalų komitetui ir Regionų komitetui – Europos pirmaujančios rinkos iniciatyva {SEK(2007) 1729} {SEK(2007) 1730}. http://eur-lex.europa.eu/legal-content/LT/TXT/?uri=CELEX:52007DC0860 (08.12.2020).

Kvieskaitė-Celešienė, E. (2011). Expression of Interfaces of Intercultural Communication and Educational Innovations: Doctoral thesis: social sciences, educology (07 S)/Eglė Kvieskaitė; Vilnius Pedagogical University. Vilnius: Publishing house of Vilnius Pedagogical University, 133 p.: iliustr. (The thesis has been prepared in Vilnius Pedagogical University in 2006–2011.) – Doctoral thesis (social sciences). Vilnius: Vilnius Pedagogical University [Tarpkultūrinės komunikacijos ir edukacinių inovacijų sąsajų raiška: daktaro disertacija: socialiniai mokslai, edukologija (07 S)/ Eglė Kvieskaitė; Vilniaus pedagoginis universitetas. Vilnius: Vilniaus pedagoginio universiteto leidykla, 133 p.: iliustr. (Disertacija rengta 2006–2011 metais Vilniaus pedagoginiame universitete.) Dr. disert. (socialiniai mokslai). Vilnius: Vilniaus pedagoginis universitetas.

Kvieskienė, G. (2013). New phenomenon: sprouts of socioeconomics between private and public sector. Future business. (Nr.1). Science popularization, educational journal. Social liability: challenges and opportunities. http://www.svak.lt/images/ateities_verslas-spauda_ok_web.pdf (08.12.2020).

Kvieskienė, G. (2016). Viešojo, privataus ir NVO sektoriaus partnerystė: tyrimo pristatymas Tyrimo ataskaita. Social Education/ The Community-Oriented Smart Socialization, t. 43, Nr. 2, pp. 78–89 / Vol. 43, No. 2, pp. 78–89. https://www.researchgate.net/profile/Giedre_Kvieskiene/publication/316019382_Public_private_ang_NGO_sectors_partnership_presentation_of_the_study/links/5bab22ec45851574f7e64f0d/Public-private-ang-NGO-sectors-partnership-presentation-of-the-study.pdf?origin=publication_detail

Kvieskienė G. & Kvieska V. (2012). Socialinės ekonomikos inovacijų įtaka visuomenės gerovei. Socialinis ugdymas, Nr. 21 (32). Vilnius: Edukologija, pp. 5–16. ISSN 1392-9569. MAKETAS_Monografija17x242012.indd (elaba.lt).

Kvieskienė, G. & Kvieska, V. (2019). Social Capital for Smart and Sustainable Development in Lithuania. https://www.socialnet.de/international/en/papers/social-capital-for-smart-andsustainable-development-in-lithuania.html (08.12.2020).

Laimingiausi žmonės gyvena Norvegijoje. http://tiesa.com/naujienos-ir-zinios/pasaulis/laimingiausi-zmones-|gyvena-norvegijoje-26709 (08.12.2020).

Matthies, A.-L., Stamm, I., Hirvilammi, T. & Närhi, K. (2019). Ecosocial Innovations and Their Capacity to Integrate Ecological, Economic and Social Sustainability Transition. *Sustainability, 11* (7), 2107. https://doi.org/10.3390/su11072107 (07.12.2020).

McKeown, R. (2002). Education for Sustainable Development Toolkit version 2. Tennessee: Waste Management Research and Education Institution. Singapore: Ministry of Education.

Milijardinė dalinimosi ekonomika – ir pavojus įprastam verslui, ir galimybės (2016). Birželio 08. http://www.veidas.lt/milijardine-dalinimosi-ekonomika-%E2%80%93-ir-pavojus-iprastam-verslui-ir-galimybes (07.12.2020).

Ministry of Education (2017). Specialusis ugdymas(is), mokymosi ir švietimo pagalba. December 6. https://www.smm.lt /web/lt /smm-svi e t imas /spec ialus is-ugdymas ismokymosi-ir-svietimo-pagalba (07.12.2020).

Ministry of Social Security and Labor (2018). Globos namų pertvarkos patirtys ir iššūkiai. https://socmin.lrv.lt/lt/naujienos/globos-namu-pertvarkos-patirtys-ir-issukiai (07.12.2020).

Misiūnienė, G. (2014). Daugiafunkcio centro socioedukacinės veiklos modeliavimas kaimo (teritorinėje bendruomenėje): Daktaro disertacija: socialiniai mokslai, edukologija (07 S)/Giedrė Misiūnienė. Lietuvos edukologijos universitetas. Vilnius: Edukologija, 229 p.: iliustr. (Disertacija rengta 2010–2014 m.) Lietuvos edukologijos universitete. Daktaro disertacija (socialiniai mokslai) Vilnius: Lietuvos edukologijos universitetas.

Mobile Garden. https://socialworkin40objects.com/2016/07/17/mobile-garden/ (08.12.2020).

Model of social partnership of public and NGO sectors (2014). Access through the Internet: http://www.nvovaikamskonfederacija.lt/wp-content/uploads/2016/03/Partnerystes-modelis-Copy.pdf (08.12.2020).

Nefas, S. (2007). Funkcionali vietos bendruomenė Lietuvos kaimuose ir miesteliuose. Daktaro disertacija. Socialiniai mokslai, vadyba ir administravimas (S03). Vilnius: Mykolo Romerio universitetas.

Neubeck K.J. (2007). Social Problems: A Critical Approach. Boston (Mass.): McGraw-Hill: Higher Education Noddings, N. Phylosophy of Education. Boulder (Colo.): Westview Press.

NGO beehive. http://www.nvoavilys.lt/ (08.12.2020).

Nijkamp., P. (2016). The «resourceful region». A new conceptualization of regional development strategies. http://old.aecr.org/images/ImatgesArticles/2016/12/09_Nijkamp.pdf (08.12.2020).

OECD (2001). The Wellbeing of Nations. The Role of Human and Social Capital. Paris OECD: Centre français d'exploitation du droit de copie. http://www.oecd.org/site/worldforum/33703702.pdf (08.12.2020).

OECD (2018). Economic Surveys Lithuania (June). http://www.oecd.org/eco/surveys/Lithuania-2018-OECD-economicsurvey-overview.pdf (08.12.2020).

Parissaki, M. & Humphreys, E. (2005). Regional Social Capital in Europe. European Foundation for the Improvement of Living and Working Conditions. Luxembourg: Office for Official Publications of the European Communities.

Partnership of public, private and NGO sector. Research report (2014). http://www.esparama.lt/documents/10157/490675/Tyrimas+NVO+akademija.pdf/259b08bc-4236-455c-a315-169b23e4ef42 (08.12.2020).

Pasaulio ekonomikos forumas (2017). https://www.weforum.org/press/2017/03/world-economic-forum-europeaninvestment-bank-urge-eu-finance-ministers-to-make-inclusive-growthtop-priority (08.12.2020).

Plot Program (2020). Vi søker nye medlemmer til Plot-rådet for 2021. 9.12 2020. www.plotoslo.no (08.12.2020).

Poulsen, M., Introduction to game based learning. The GameIt. A framework of Game based pedagogy. Edited by Mathias Poulsen & Ebba Køber. The GameIt handbook by Fleksibel utdanning Norge – issuu. http://www.mathiaspoulsen.com/gemeithandbook.pdf (08.12.2020).

Poulsen, M. & Køber, E. (eds.) (2011). The GAMEiT Handbook. A Framework of Game Based Learning Pedagogy. Oslo: Play Consulting GameIT College. http://www.mathiaspoulsen.com/gemeithandbook.pdf (07.12.2020).

Positive socialisation, entrepreneurship and social responsibility. Vilnius (2014) – spalio 27–28 d. Vardas, pavardė Simpoziumo globėjai: Lietuvos Respublikos Seimo Švietimo mokslo ir kultūros komitetas/ Lietuvos Respublikos Seimo Socialinių reikalų ir darbo komitetas. http://www3.lrs.lt/docs2/DHXNYELG.PDF (08.12.2020).

Promotion of sustainable development by strengthening social clustering of public sector: http://www.socialcluster.lt/ (08.12.2020).

Ralys, K. (2011). Preparation for Marriage and Family: Scientific Monograph; Vilnius: Publishing house of Vilnius Pedagogical University.

Reforms in Lithuania are reinforcing economic growth but boosting productivity is still a challenge. http://www.oecd.org/economy/reforms-in-lithuania-are-reinforcing-economicgrowth-but-boosting-productivity-is-still-a-challenge.htm (08.12.2020).

Reid, A. et al. (2008). Participation and Learning: Perspectives on Education and the Environment, Health and Sustainability. Dortrecht: Springer School Re-Design & New School Innovation. http://www.educationresourcesconsortium.org/school-re-design (08.12.2020).

School Re-Design & New School Inovation. http://www.educationresourcesconsortium.org/school-re-design/ (08.12.2020).

Seven things to know about Lithuania's economy. A crash-course on the Lithuania neconomy, which becomes part of the Eurozone on 1 January. https://www.politico.eu/article/lithuanias-economy-seven-facts-to-introduce-you/ (08.12.2020).

Smart Education in Smart Cities and Smart Regions (2017). New Education Forum. https://www.neweducationforum.com/documents/Report-NEF-Brussels2017.pdf (08.12.2020).

Social Capital. https://www.oecd.org/insights/37966934.pdf (08.12.2020).

Social Life and Social Knowledge – Toward a Process Account of Development (2008). New York (N.Y.); London: Lawrence Erlbaum Associates – Taylor &Francis Group.

Socialinis ugdymas. https://www.smm.lt/uploads/documents/svietimas/ugdymo-programos/6_Socialinisugdymas.Pdf (08.12.2020).

Socialinių verslininkų interneto portalas. www.socialinisverslas.lt (08.12.2020).

SPPD (2018). Licencijos. http://www.sppd.lt/lt/veiklos-sritys/licencijavimas/licencijos/ (08.12.2020).

Stenros, J. (2014). In: Defense of a Magic Circle: The Social, Mental and Cultural Boundaries of Play. http://todigra.org/index.php/todigra/article/view/10/26 (07.12.2020).

Stroputė, N. & Kairytė., M. (2015). How social business becomes social business? Analysis of good foreign practice. http://kurklt.lt/wp-content/uploads/2015/11/U%C5%BEsienio-%C5%A1ali%C5%B3-SV-vystymosiir-pl%C4%97tros-gerosios-praktikos-analiz%C4%97_Neringa-ir-Modesta_2016_Kurk-Lietuvai_.pdf (08.12.2020).

Šalkauskis, S. (1992). Visuomeninis auklėjimas. In: Rinktiniai raštai. II knyga. Vilnius: leidybos centras.

The Children's Art Club Rasmus. http://afmuseet.no/en/besak-oss/barnekunstklubben-rasmus (08.12.2020).

The Legatum Prosperity Index™ (2016). http://www.prosperity.com/ (08.12.2020).

The Mar City Market (2013). The Smart City Market: Opportunities for the UK. Department for Business, Innovation and Skills 1 Victoria Street London SW1H 0ET. www.gov.uk/bis. Research paper number 136 October. https://assets.publishing.service.gov.uk/government/uploads/system/uploads/attachment_data/file/249423/bis-13-1217-smart-city-market-opportunties-uk.pdf (08.12.2020).

The Minister of Economy of the Republic of Lithuania (2015). The Order regarding amendment of the Order No. 4–207 of the Minister of Economy of the Republic of Lithuania: "Regarding approval of the Conception of social business", dated on the 3rd of April. https://ukmin.lrv.lt/uploads/ukmin/documents/files/SVV/Soc_%20v_%20koncepcijos%20pakeitimas%202016-08-29.pdf (08.12.2020).

The Unido ApproAch to Business Investment Technology Services Key Principles and Project Experiences for Inclusive Growth Cluster Development (2014). https://www.unido.org/sites/default/files/2014-01/

UNIDOs_CLUSTER_APPROACH_0.PDF (08.12.2020) and: https://www.unido.org/fileadmin/user_media_upgrade/What_we_do/Topics/Business__investment_and_technology_ (08.12.2020).

The world's happiest, safest and richest countries. http://www.telegraph.co.uk/travel/news/prosperityindex-2016-happiest-countries (08.12.2020).

Torres, A.C. & Antikainen, A. (2003). The International Handbook on the Sociology and Education: An International Assesment of New Research and Theory. Boston: Rowman and Littefield Publishing groups.

TUNÇ KARADAĞ (2013). In: Partial Fulfillment of the Requirements for the Degree of Master of Science in City Planning. http://citeseerx.ist.psu.edu/viewdoc/download?doi=10.1.1.633.3072&rep=rep1&type=p (08.12.2020).

Tvarios plėtros skatinimas stiprinant viešojo sektoriaus socialinę klasterystę: http://www.socialcluster.lt – and: http://www.socialcluster.lt/en/kontaktai/ (08.12.2020)

Vyšniauskaitė, M., Kvieskienė, G., Kvieska, V. & Rauhansalo, T. (2017). Kaip ugdyti sąmoningą moksleivį. Integruoto ugdymo gairės. http://www.llri.lt/wp-content/uploads/2017/02/kaip-ugdyti-socialiaisamoninga-moksleivi.pdf (08.12.2020).

Wehr, H. & Carlsburg, G.-B. v. (Hrsg.) (2020). Erlebnispädagogik. Theorie, Praxis und Projekte für die Schule (= Pädagogik und Psychologie). 5. Auflage. Augsburg: Brigg.

Welfare (Legatum) index. https://www.prosperity.com/rankings (08.12.2020).

Žadeikaitė, L. & Gulbinas, R. Socialinio pedagogo socioedukacinės aplinkos tobulinimas panaudojant IKT. http://socialinisugdymasleu.lt/index.php/socialinisugdymas/article/view/38/34 (08.12.2020).

Zaleskienė, I. (2009). Jaunimo pilietiškumas: tarp teksto ir konteksto. Tautiškumas ir pilietiškumas. Atskirtis ar dermė? Habilitacijos procedūrai teikiamų mokslo darbų apžvalga. Socialiniai mokslai, edukologija (07 S). Vilnius: Vilniaus pedagoginis universitetas.

About the Authors

Giedrė Kvieskienė

Prof. PhD., Giedrė Kvieskienė is an expert in social economy and education policy, with a focus on positive socialization, communication and social business. Her research interests relate in particular to social partnership, social business and smart socialization. She was the project coordinator for EU-Tempus project in Lithuania, advancing the three-cycle system of social work education in six European countries. Prof. Dr. G. Kvieskienė was the member of the European Consul program, based on the research on policies and practices teaching socio-cultural diversity (2006–2009). She is also Honor President NGO Confederation for Children, Expert of the European Anti-Poverty Network and professor of the Social Education Faculty in Lithuanian University of Educational Science and Lithuanian CIVITAS INTERNATIONAL coordinator in the "CIVITAS INTERNATIONAL" world network. She is a member of a Socrates project, New Modules of Activism in European Social Work and creator of a module on cultural diversity and migration.

Vytautas Kvieska

Innovator and Lecture of Social Economics Kvieska teaches social economics, social business and couching modules for baccalaureate, and social economic and entrepreneurship modules for MA degree students in Lithuanian Education Science and Vytautas Magnus Universities.

His research interests are: social economics, marketing, innovation, social partnership and project management. He published monograph, methodical material and more than 30 articles in scientific, practical fields. He is also a pioneer of start-ups in innovative lighting projects, and one of the first authors of articles about social economy and one of the authors of widely known monograph on social partnership and innovation.

Lecture of Vytautas Magnus University Vytautas Kvieska teaches courses on Social Economy, Couching and Social Industries for Bachelors, Social Economy and Entrepreneurship for Master's degree students.

Gerd-Bodo von Carlsburg

Prof. Dr. Dr. h.c. mult., he completed a teaching degree at the University of Hamburg and passed here in 1970 the First State Examination, 1972, the Second State Examination. He taught at the Heinrich-Hertz-Schule in Hamburg, mainly at the upper level of the Abt. Gymnasium. In 1972, he received his doctorate and worked as a lecturer at the University of Hamburg. In parallel, from 1973 to 1975, he worked as an Academic Counselor at the University of Wuppertal; from March 1, 1975, as a professor at the University of Education Heidelberg. From 1982 to 1984, he was a representative of the Department of History and Systematics of Education at the Helmut Schmidt University of Hamburg. From 1984 to 1987, he took on teaching duties at various universities in addition to his teaching duties. From 1991 to 1994, he was head of faculty at the Heidelberg University of Education and vice-dean in 1994/95, later director of the institute. From 1994, he also took a visiting professorship at the Vilnius Pedagogical University, later Lithuanian University of Education, and received an honorary doctorate in 1995. From 1996 to 2000, he was also a visiting professor at Tallinn University of Education, now Tallinn University, and was awarded an honorary doctorate in 1999. After 2000 he was also guest professor at the Academy of Music Tallinn, until 2012 and 2016 as well as 2018 to 2020 lecturer at the University of Education Karlsruhe and until 2012 also lecturer at the University of Music Trossingen, 1992, appointed as Colonel (reserve).

In addition, for two years he was chairman of the DGfE Section for Differential Educational Research and the DGfE Psychoanalytical Education

Commission (DGFE - German Association for Educational Science). He is President of the Weltbund für Erneuerung der Erziehung (WEE) e. V. (World Education Fellowship - German Language Section), honorary member of the EIPOS Institute e.V. (European Institute for postgraduate Education at the Technical University of Dresden) and member of the National Board of Trustees of the CJD e. V. (Das Christliche Jugenddorfwerk Deutschlands e.V.).

Baltische Studien zur Erziehungs- und Sozialwissenschaft

Baltic Studies in Educational and Social Sciences

Herausgegeben von Gerd-Bodo von Carlsburg, Natalija Mažeikienė und Airi Liimets

Edited by Gerd-Bodo von Carlsburg, Natalija Mažeikienė and Airi Liimets

Band	1	Gerd-Bodo Reinert / Irena Musteikienė (Hrsg.): Litauische Gespräche zur Pädagogik. Humanismus - Demokratie - Erziehung. 1999.
Band	2	Gerd-Bodo Reinert / Irena Musteikienė (Hrsg.): Litauische Gespräche zur Pädagogik II. Staat und Schule. 2000.
Band	3	Jaan Mikk: Textbook: Research and Writing. 2000.
Band	4	Gerd-Bodo Reinert / Irena Musteikienė (Hrsg.): Erziehungswandel und moderne pädagogische Verfahren. 2001.
Band	5	Airi Liimets (Hrsg.): Integration als Problem in der Erziehungswissenschaft. 2001.
Band	6	Gerd-Bodo Reinert / Irena Musteikienė (Hrsg.): Wissenschaft - Studium - Schule auf neuen Wegen. 2002.
Band	7	Palmira Jucevičienė / Gediminas Merkys / Gerd-Bodo Reinert: Towards the Learning Society. Educational Issues. 2002.
Band	8	Gerd-Bodo Reinert / Irena Musteikienė (Hrsg.): Bildung im Zeitalter der Informationsgesellschaft. 2003.
Band	9	Dalia Ohlinger: Argumentation in der Erst- und Fremdsprache. Pragmalinguistische und grammatikalische Aspekte anhand von Argumentationen deutscher und litauischer Studierender. 2003.
Band	10	Inga Šukevičiūtė: Interkulturelle Interferenzen im Bereich Wirtschaftskommunikation. Zur Analyse von Geschäftsbriefen im Vergleich Deutschland, Litauen und Russland. 2004.
Band	11	Gerd-Bodo Reinert von Carlsburg / Irena Musteikienė (Hrsg./eds.): Innovation durch Bildung. Innovation by Education. 2004.
Band	12	Gerd-Bodo von Carlsburg / Palmira Jucevičienė / Gediminas Merkys (eds.): Learning and Development for Innovation, Networking and Cohesion. 2005.
Band	13	Gerd-Bodo von Carlsburg / Irena Musteikienė (Hrsg./eds.): Bildungsreform als Lebensreform. Educational Systems Development as Development of Human Being. 2005.
Band	14	Gerd-Bodo von Carlsburg (Hrsg./ed.): Entwicklung erziehungswissenschaftlicher Paradigmen: Theorie und Praxis. Development of Educational Paradigms: Theory and Practice. 2007.
Band	15	Gerd-Bodo von Carlsburg (Hrsg./ed.): Bildungs- und Kulturmanagement. The Management of Education and Culture. 2008.
Band	16	Jaan Mikk / Marika Veisson / Piret Luik (eds.): Reforms and Innovations in Estonian Education. 2008.

Band 17 Gerd-Bodo von Carlsburg (Hrsg./ed.): Qualität von Bildung und Kultur. Theorie und Praxis. The Quality of Education and Culture. Theoretical and Practical Dimensions. 2009.

Band 18 Airi Liimets (Hrsg.): Denkkulturen. Selbstwerdung des Menschen. Erziehungskulturen. Festschrift für Professor Dr. Dr. h.c. Dr. h.c. Heino Liimets. 2010.

Band 19 Ellu Saar (ed.): Towards a Normal Stratification Order. Actual and Perceived Social Stratification in Post-Socialist Estonia. 2011.

Band 20 Marika Veisson / Eeva Hujala / Peter K. Smith/ Manjula Waniganayake / Eve Kikas (eds.): Global Perspectives in Early Childhood Education. Diversity, Challenges and Possibilities. 2011.

Band 21 Airi Liimets / Marit Mäesalu (eds.): Music Inside and Outside the School. 2011.

Band 22 Gerd-Bodo von Carlsburg (Hrsg./ed.): Enkulturation durch sozialen Kompetenzerwerb. Enculturation by Acquiring of Social Competences. 2011.

Band 23 Reet Liimets: Ich als raumzeitliches Konstrukt. Die Fiktionen vom Leben der estnischen und deutschen Jugendlichen. 2012.

Band 24 Raivo Vetik (ed.): Nation-Building in the Context of Post-Communist Transformation and Globalization. The Case of Estonia. 2012.

Band 25 Airi-Alina Allaste (ed.): ‚Back in the West'. Changing Lifestyles in Transforming Societies. 2013.

Band 26 Gerd-Bodo von Carlsburg (Hrsg./ed.): Bildungswissenschaft auf der Suche nach globaler Identität. Educational Sciences in Search of Global Identity. 2013.

Band 27 Airi Liimets / Marika Veisson (eds): Teachers and Youth in Educational Reality. 2014.

Band 28 Gerd-Bodo von Carlsburg / Thomas Vogel (Hrsg./eds.): Bildungswissenschaften und akademisches Selbstverständnis in einer globalisierten Welt. Education and Academic Self-Concept in the Globalized World. 2014.

Band 29 Marika Veisson / Airi Liimets / Pertti Kansanen / Edgar Krull (eds.): Tradition and Innovation in Education. 2015.

Band 30 Robertas Jucevičius / Jurgita Bruneckienė / Gerd-Bodo von Carlsburg (eds.): International Practices of Smart Development. 2015.

Band 31 Gerd-Bodo von Carlsburg (Hrsg./ed.): Strategien der Lehrerbildung. Zur Steigerung von Lehrkompetenzen und Unterrichtsqualität. Strategies for Teacher Training. Concepts for Improving Skills and Quality of Teaching. 2016.

Band 32 Gerd-Bodo von Carlsburg (Hrsg./ed.): Denk- und Lernkulturen im wissenschaftlichen Diskurs. Cultures of Thinking and Learning in the Scientific Discourse. 2017.

Band 33 Palmira Pečiuliauskienė / Aleksa Valdemaras: Motivation of New Generation Students for Learning Physics and Mathematics. 2018.

Band 34 Gerd-Bodo von Carlsburg (Hrsg./ed.): Transkulturelle Perspektiven in der Bildung. Transcultural Perspectives in Education. 2019.

Band 35 Giedrė Kvieskienė / Vytautas Kvieska / Gerd-Bodo von Carlsburg: Social Clustering: Paradigm of Trust. 2021.

www.peterlang.com

www.ingramcontent.com/pod-product-compliance
Ingram Content Group UK Ltd.
Pitfield, Milton Keynes, MK11 3LW, UK
UKHW041825031026
15587UKWH00002B/564